THE LAST EXTINCTION

THE LAST EXTINCTION

The REAL SCIENCE *Behind the* DEATH *of the* DINOSAURS

Gerta Keller, PhD

DIVERSION
BOOKS

Diversion Books
A division of Diversion Publishing Corp.
www.diversionbooks.com

Copyright © 2025 by Gerta Keller

All rights reserved, including the right to reproduce this book or portions thereof in any form whatsoever. No part of this publication may be reproduced or transmitted in any form or by any means, electronic or mechanical, including photocopying, recording, or any other information storage and retrieval, without the written permission of the publisher.

Diversion Books and colophon are registered trademarks of Diversion Publishing Corp.

For more information, email info@diversionbooks.com.

Hardcover ISBN: 979-8-89515-046-7
e-ISBN: 979-8-89515-047-4
First Diversion Books Edition: September 2025

Design by Neuwirth & Associates, Inc.
Cover design by Jonathan Sainsbury / 6x9 Design

Printed in the United States of America
1 3 5 7 9 10 8 6 4 2

Diversion books are available at special discounts for bulk purchases in the US by corporations, institutions, and other organizations. For more information, please contact admin@diversionbooks.com.

The publisher does not have any control over and does not assume any responsibility for author or third-party websites or their content.

CONTENTS

The Nastiest Feud in Science		vii
1	The Bomb	1
2	The Search for Truth	13
3	In the Crosshairs	29
4	Snowbird	47
5	The Girl Who Didn't Know Her Place	57
6	The Crater	73
7	Under the Revival Tent	87
8	Crazy Fun (Six Years of Detective Work)	103
9	The Mysterious Case of the Vanishing Cores	119
10	Race to the Deadline	127
11	Showdown in Nice	141
12	Time to Be Bold	155
13	"Oohing and Aahing Over Rocks"	167
14	Turning the Tide	189
15	An Outrageous Idea	201
16	The Truth About Chicxulub	217
17	Explaining the "Iridium Anomaly" (Confession of a Former Impactor)	231
18	Are We Living in the Sixth Extinction?	241
Acknowledgments		245
Notes		251
About the Author		255

THE NASTIEST FEUD IN SCIENCE

Introduced as "the well-known sedimentologist George Keller," I walked onto the stage to snickering laughter. At the podium I adjusted the microphone, took a deep breath, and said, "I am Gerta Keller, the *paleontologist.*" More laughter came from the sea of grinning male faces before me. The auditorium was packed to overflow, with people standing in the back and along the walls. These were the heavy hitters of my field—the world's leading geochemists, geophysicists, paleontologists, and other scientists—gathered at the prestigious 1988 Snowbird II conference on Impacts and Mass Extinctions. I quieted the voice in my mind that told me I didn't belong here; my work would speak for itself.

The title of my presentation flashed across the screen behind me: *Chicxulub—the Non-Smoking Gun: Impact Crater Predates K–P Boundary,* and the laughter turned to murmurs of consternation and scorn. I began my talk, methodically walking the audience through the data I'd carefully assembled over the past three years. My nervousness turned into excitement as I made my case. I'd waited a long time for this moment. I knew my findings had the potential to redirect the entire conversation around one of the great mysteries in my field—what caused mass extinctions.

Only minutes after I'd begun my presentation, a tall, imposing man rose from the audience and strode to the microphone positioned in the aisle in the middle of the auditorium. I knew him: Jan Smit. Smit was assistant professor of paleontology and geology from the University of Utrecht and a fervent disciple of the impact mass extinction theory in its most dramatic version: A giant meteorite ten kilometers in diameter crashed into Earth, triggering freezing darkness, global wildfires, earthquakes, and mega-tsunami waves that devastated life on land and sea. Smit stared me down, a smug smile playing at his lips. The energy in the

viii THE NASTIEST FEUD IN SCIENCE

room changed, charged with nervous anticipation. The crowd registered it before I did: A fight was brewing.

As I continued my talk, doing my best to ignore Smit, other scientists lined up behind him and the two other microphones positioned in the side aisles. One by one they rose until dozens of men, tense and visibly angry, stood in silence. The microphones were to be used for follow-up questions, discussion, and critique, but as I watched the men heave with outrage, one almost purple in the face, it became clear that they were to be the weapons of my public humiliation.

I was only halfway through my presentation when Smit interrupted in a booming voice, "You are simply wrong! Everyone knows the mass extinction was instantaneous and caused by the impact."

A chorus of shouts rose in support.

You know nothing!

Garbage!

You're ignorant!

A sick despair swept through me. I paused to collect myself and took a drink of water. My hand shook as I raised the glass. I pressed on, pointing out the data that defended my claims, but was quickly shouted down again.

You don't know what you're doing!

Stupid!

For the next forty-five minutes, a litany of insults rained down on me as some of the world's most renowned geophysicists, planetologists, and astrophysicists took turns publicly insulting me. Repeatedly, they asked me, *Why does your data not confirm Jan Smit's mass extinction theory?* There was no interest in the science, no questions about my results—just insults, denials, threats, and rage against an upstart, a woman no less, who dared to question their theory. Relishing his turn at the microphone, Smit shouted the loudest, attacking my data fiercely, though without substance. I could see his goal was to intimidate and isolate me, and ultimately to shut me down.

My cheeks burned and my heart threatened to gallop outside of my chest. I was confident in my results and knew I could ably defend them if given the opportunity, but I couldn't get a word in above the constant

THE NASTIEST FEUD IN SCIENCE   ix

shouting. Never did I imagine this type of bullying behavior could occur at a professional conference.

Weren't scientists supposed to be objective in their pursuit of the truth? If so, why did these men so passionately deny my research before they had even heard it? Why were they attempting to silence me? If they weren't interested in contending with the evidence I was presenting, what did they want? I would wrestle with these questions for years to come. All I knew in the moment, shaking onstage, was that I was telling the truth according to the facts. If my research was right, then an asteroid did not wipe out the dinosaurs.

My attempt to share my research at the Snowbird II conference was the beginning of what has become popularly known as "The Dinosaur Wars," a contentious debate over what triggered the fifth mass extinction at the end of the Cretaceous Era sixty-six million years ago: in other words, what really killed the dinosaurs. Our real-world monsters, dinosaurs have enthralled us for generations, and the question of what caused their demise is more relevant than ever as humankind confronts the paroxysms of an imperiled planet and the possibility that *we* may become the dinosaurs of the sixth extinction.

It's been called the nastiest feud in science. Passions in this field run deep. Most people don't think of science as a blood sport; this book may change that. *The Last Extinction* is the story of the forty-three years I've spent in the arena, fighting—and winning—against a toxic scientific establishment determined to bury me, all for my stubborn pursuit of the truth. Because I dared to challenge the popular asteroid impact theory, Smit's contingent launched an all-out war against me, doing their utmost to sabotage my work, destroy my reputation, and suppress the publications of my research. I've been told a lesser individual might have cut and run, but quitting was never an option for me. Smit picked a fight with the wrong woman.

To succeed, I've had to harness all my courage and resilience, forged over a lifetime of unconventional experiences. I've had to be a fighter, and I hope to show other women how they, too, can fight for their careers in science.

This is the story of how I and a small team of collaborators were finally able to share our exhilarating discoveries in the face of a unified

x   THE NASTIEST FEUD IN SCIENCE

group of establishment scientists who were unwilling to listen to reason. In it you will find betrayal and sabotage, as well as perseverance, vindication, and the power of truth, and woven through it all, the dinosaurs, impact, and volcanism. It is, from where I sit, the greatest scientific detective story of our time.

THE LAST
EXTINCTION

1

The BOMB

I MADE MY FIRST GROUNDBREAKING DISCOVERY IN 1977 WHILE still a graduate student at Stanford University. It was aboard *Glomar Challenger*, a large ocean-drilling ship funded by the National Science Foundation (or NSF). For six weeks, our research crew bored through and analyzed sediments from the Japan Trench, located off northeastern Japan. We were a floating island, crusted in salt and isolated in rough seas with no land in sight. I was the only woman among the twelve scientists living in close quarters.

The Japan Trench is part of the Pacific Ring of Fire, a belt of volatile geological activity that girdles most of the Pacific Ocean. Created by the interaction of many different tectonic plates, the ring is the site of two-thirds of the planet's volcanoes and 90 percent of our earthquakes. Most of these events are caused by subduction, when an oceanic plate collides with a continental plate, and one dips below the other. This still-little-understood process was our focus on this mission. Using *Glomar Challenger*'s large drill, the scientists on board were probing into the deepest part of the trench at 26,398 feet (8,046 meters) to learn more about subduction zones.

Glomar Challenger was a great ship, painted white and navy and squeaky clean in appearance, except for the giant rust-corroded drilling rig positioned in the ship's center. When I moved between the ship's

2   THE LAST EXTINCTION

interior state-of-the-art labs and the hulking drill rig on the deck, I felt like I traversing two worlds, one modern and one industrial. The drilling operation continued around the clock in two shifts, and so did the science crews. We were two teams, one Japanese and one American, each overseen by a respective co-chief. We worked during the turbulent stormy season. I recall being perpetually cold and smattered with salt, but I relished the experience.

The drilling operation on *Glomar Challenger* fascinated me. I was amazed at how far the ship's drill could reach. The area we were studying is so deep that if you tipped Mt. Everest upside down into the ocean, its peak would still fall 1,150 feet (350 meters) short of grazing the seafloor. I would think about this as we floated on the surface, and the ship's diamond-studded drill bit bored deeper and deeper through the rocks on the ocean floor. Every one to two hours, the drill would bring core samples, ten meters long and six centimeters wide, back to the surface. Then the scientists would get to work. In the lab we cut these cores into one-and-a-half meter-long segments and then sliced them in half lengthwise. Each core segment half was placed into two core boxes. One core box was saved for the core archive library as a record for posterity. The other "working half" was studied onboard.

My task was to age-date the sediment samples by analyzing the microfossils found within them. It was an ideal role for me; my advisor at Stanford had steered me into quantitative foraminifer (foram for short) research in Pacific deep-sea sites as part of my thesis project. Each day I processed the core samples for these fossils to study under a microscope. The crew and I expected that as we drilled deeper into the trench, the microfossils we found would confirm the successively deeper and older age of the sediments.

But evidence doesn't always cooperate with our expectations. One day, instead of finding progressively older deep-sea forams in the sediment layers, as one would predict, we discovered millimeter-sized benthic forams in one of our trench samples. This made no sense. Benthic forams lived and scavenged for food on beaches. There was no way a benthic could have lived at a depth of 4,000 meters. It was like finding a crocodile at the top of the Empire State Building. I lifted my eyes from the

THE BOMB 3

microscope and shook my head to clear my mind from this hallucination. How could organisms commonly found at shorelines end up in a deep ocean trench? I peered into the scope again, and there they were—and not just a few, but a complete assemblage of many thriving species you'd normally find on a beach. Heading over to the ship library, I researched the depositional depths at which such assemblages could be found and confirmed it was a beach foram assemblage.

My heart raced as I examined the forams under the microscope, confirming again that what I'd glimpsed was real. But how did it get into the deepest part of the ocean? There was only one reasonable interpretation: We had found the fauna in what was once a beach of a small island that was located on a descending plate that was 'swallowed' in the subduction zone of the Japan Trench. It was a new and revolutionary discovery—island fauna had never been observed in any subduction zone.

Piecing together this mystery gave me an adrenaline rush akin to the runner's high but exponentially more thrilling. This feeling was one reason I'd become a scientist. I couldn't wait to share these findings with my colleagues. I assumed they'd also be excited by the fascinating story that the evidence we'd discovered, in the form of little fossils collected from the distant ocean floor, was telling us. But new discoveries are not always well received by the scientific establishment, especially if they challenge accepted beliefs. Even more so, I was to learn, if they come from a woman.

The blowback from my colleagues was immediate. The Japanese co-chief and his scientists furiously disagreed with my unorthodox conclusion. It did not conform with their expectations. They had hypothesized that the age of the sediments would simply grow older as the drill dug deeper. Now I was telling them an island was riding piggyback on the seafloor that was sucked beneath the colliding continental plate.

One by one the Japanese scientists visited me in the lab and attempted to persuade me that I was wrong. The first one told me that although I was "smart for a woman scientist," I "must change my mind."

"I can't change my mind because the evidence is here," I said, pointing at the forams under the microscope. "Do you want to see it?"

He shook his head vigorously.

4   THE LAST EXTINCTION

My second visitor told me I was "acting like a man" by not standing down. He seemed as upset by the fact that I, a woman, had found the evidence as he was by the evidence itself. Our conversations ended with him shouting "Women are not scientists!" before storming out of the lab. Then came the last visitor, whose visit was much more ominous. After I repeated my position, he turned his back on me and, gazing out at the roiling ocean, told me that if I didn't change my mind "bad things will happen to you."

Had I misunderstood him? "And what will that be?" I challenged.

"You may drown," he said. He turned to look me in the eye. "Accidents happen," he said simply, and walked out.

My stomach turned. I stared at my hands, the microscope, my desk. Had he just threatened my life? Over scientific evidence? I simply couldn't understand what could motivate him to make such a threat. It was so ridiculous; I wanted to laugh. But even if this was just some aggressive posturing on his part, I also realized how easily they could dispatch me if they really wanted to. There were plenty of opportunities for them to throw me overboard. Someone could knock me off the aft during my daily runs or toss me over the railing during the night shift when I was working alone. There would be few if any witnesses.

The threat of male violence wasn't new to me. Growing up in Switzerland in the mid-twentieth century meant coming of age in a culture that conditioned women to obey men. Those who stepped out of line were ostracized or worse. Even women who conformed to strict standards of Swiss propriety experienced frequent, undisguised sexual harassment. As a teenager, I had worked clearing tables and cleaning dishes in restaurants during the weekends; catcalling, grabbing, and groping were all too common. As I matured in Switzerland, harassment progressed to sexual assaults, which continued after I arrived in the States and attended college. I survived these attacks, though for years after lived in fear of the next one.

Now, years later, alone in the ship's lab, I felt that familiar fear. How far would the Japanese scientists go to silence me? For a moment, I wavered. Should I give in? Was it worth my life? Then my senses kicked in: I hadn't come this far in life to let prickly male egos rob me of a major discovery.

The next day, the US co-chief, Roland von Huene, visited me in the lab. Rollie was a jovial character, with a shock of red hair and a face crisscrossed with laugh lines. I saw him as a true scientist and trusted him, but I was wary of this visit. Would he side with the Japanese co-chief and try to convince me to throw out my findings for the sake of peace?

He sat down on my desk and broke out in an ear-to-ear grin. "Gerta, you've created some war!" he said with delight. "The Japanese are ready to throw you overboard unless you retract your scenario of the subducted island. It's a good story." He leaned in. "But now tell me, how certain are you with your evidence?"

I sighed deeply and told Rollie, "The evidence is real. You can't deposit a beach foram assemblage in the Japan Trench without transporting it there in one piece and that can only be done in subduction. You are an expert on this process and know I'm right."

He stared at me for some time. "Are you really sure?" he asked again.

The note of doubt in his voice emboldened me. I plunked my coffee cup down on my desk for emphasis, sloshing coffee over my papers. "I've never been more certain," I said firmly.

Rollie put his hand on my shoulder. "Gerta," he said. "I'll stand behind you all the way. And I will deal with the co-chief. They'll leave you alone."

I must have looked over my shoulder a hundred times a day for the remainder of the voyage. The Japanese scientists never spoke to me again, except for the senior scientist, who took great delight in my discovery.

A year later the Japanese science crew published their important new discovery of a swallowed island in the Japan Trench. This story received much acclaim and publicity, including an article that ran in *The New York Times* in 1979. There was no mention of the woman scientist who made this discovery. At the time, I attributed this unhappy experience to the sexism of Japanese culture. Surely, this could not happen in the West, I thought. It didn't take long to find out how wrong I was.

The next year, I earned my PhD in climate and environmental studies focused on the analysis of microscopic single-celled forams to determine global temperature changes. My research was based on the work of Nick Shackleton of Cambridge University. He pioneered foram-based climate

6 THE LAST EXTINCTION

studies in the mid-1960s, by using carbon and oxygen isotopes in foram shells to evaluate temperature changes. This method was so successful it continues to play an important role in establishing the global climate record today.

In 1979, I decided to use the skills I had acquired studying forams to better understand mass extinction events, particularly the one that killed the dinosaurs. This was a return to a subject that had captivated me since my undergraduate years at San Francisco State University. Little was understood about this topic before the 1970s, when I was beginning my first independent analysis of vertebrate fossils (those with backbones) and invertebrate fossils (those without backbones). My research at the time was based at the California Academy of Sciences in San Francisco, which held the largest fossil collection in the state. I spent many, many hours as an undergraduate sorting through this trove of data, looking for patterns that hadn't been noticed before. When I started graduate school at Stanford, I put aside this work, along with the paper I had written on the subject. Now I could pick up where I left off, but with a powerful new tool. I knew that forams, despite their small size, could be a rich source of information about the fifth mass extinction.

At that time, I held two jobs: one as an independent researcher at Stanford University, the other as a National Research Council Fellow at the United States Geological Survey (USGS) in Menlo Park. At the USGS, I spent much of my time interacting with colleagues from different disciplines in marine geology and collaborating with them on a variety of projects. Like a sponge, I soaked up whatever I could learn from them. My partner, Andy Majda, taught mathematics at UC Berkeley and spent most of the week in an apartment there to be close to his office, which freed me to indulge my inner research fanatic. I frequently lost myself in deep dives into my work for up to eighteen hours at a time. On alternate weekends, I joined Majda in Berkeley, which became our launch point for spectacular explorations of the North Bay, including our favorite breathtaking drives to Point Reyes. I was happier and more settled than I'd ever been before. But our quiet bliss didn't last long.

On June 6, 1980, I'd already spent hours hunched over my microscope in my office in Menlo Park when I heard the news. At the time,

THE BOMB 7

I was peering at slides of forams found in a sediment layer taken from Tunisia that was dated to the time of the fifth mass extinction. It was painstaking, repetitive work that required total concentration. My neck and shoulders ached, my eyes were red from strain, and I loved every moment of it. I was a year into my concerted investigation of the cause of the fifth mass extinction. I was finally making the scholarly leap I'd dreamed of.

The fifth mass extinction occurred sixty-six million years ago and marks the transition from the Cretaceous to Paleogene periods. "Cretaceous" is derived from the Latin word for "chalk," which was abundant in the soil at that time. (The Germans call chalk "kreide," which is why Cretaceous is often abbreviated as "K".) The Paleogene, P or Pg, is the younger period that began after the mass extinction. The mass extinction boundary is labeled KPB for Cretaceous–Paleogene Boundary.[1]

The Cretaceous is of special interest to the public because it was the last period in which large dinosaurs roamed the Earth. (It is common to say the dinosaurs went extinct, but of course, not all dinosaurs died out because birds evolved from dinosaurs and thrive in great abundance to this day.) Everyone knows about dinosaurs, especially little kids. But most people don't know that the most dramatic disappearance of species occurred not on land, but in the oceans, or that the fossil record shows that the "death of the dinosaurs" didn't happen suddenly but took place over a long period time. In fact, at the time the Cretaceous period was coming to an end, and the fifth extinction was underway, dinosaur species had already been decreasing in diversity for over seven million years. This was mainly due to climate warming, the drying up of land in the Western Interior Seaway of the US, and a dramatic increase in volcanic eruptions in India.

The last surviving species of the non-avian dinosaurs are believed to have been the large land-dwelling *Tyrannosaurus rex* and its smaller African contemporary *Chenanisaurus barbaricus*, both ferocious predators. But the fifth mass extinction didn't only affect large animals like dinosaurs. In fact, the species most affected by the fifth mass extinction happened to be the tiny forams that I researched. These single-celled animals were at the opposite end of the size spectrum from the mighty dinosaurs, the largest

8   THE LAST EXTINCTION

foram being about the size of a small pinhead. Different foram species met different fates during the fifth extinction. Most benthic forams—the species I unexpectedly found in the Japan Trench—survived. But planktic forams, which thrived before the events that caused the mass extinction began, suffered near total extinction. In fact, no other fossil groups show such a dramatic mass extinction at the end of the Cretaceous.

Thus, if you look at sediment layers from before and after the fifth extinction, in locations all over the world, you will find many layers rich with planktic forams and then, suddenly, just one survivor. This is why the near absence of Cretaceous planktic forams in a sediment layer has become the established marker for the K–P catastrophe. And it's one of the reasons I was so interested in studying these tiny, beautiful, ancient creatures. I believed that foram analysis across the Cretaceous–Paleogene boundary (KPB) would yield clues to solving the mystery of what snuffed out life, including the dinosaurs.

And that's what I was doing on that June morning in 1980. Through the microscope before me, I peered at slides of tiny foram species extracted from the sediments of a large rock exposure, or outcrop, near the Tunisian city of El Kef. Like any rock outcrop useful to geologists, this one revealed a layer-cake-like sediment accumulation over tens of thousands to hundreds of thousands of years. Geologists love to collect such sediments layer by layer to study and determine the age, climate, and environmental history of a given time interval.

At the El Kef sequence, the most important feature of the KPB is a fifty-centimeter-thick black clay layer with a four-millimeter-thin red clay near the base. The thin layer of clay is iron-rich and oxidized, which is why it has a red color. Most Cretaceous species disappeared at or before the red clay, which makes it a convenient mass extinction horizon. The overlying clay is black because it contains ample amounts of organic matter that was no longer consumed after the mass extinction. Once animal life returned via evolution and normal oxygen conditions resumed, the sediments turned gray again. These red and black clay layers largely record the environment in the aftermath of the mass extinction.

As I studied my slides, I heard a growing commotion in the hallway. Curious, I poked my head outside my office door. My colleagues at the

THE BOMB   9

USGS had packed into the coffee room, and others were spilling into the hall. Everyone was talking, some in heated discussions, others laughing and shaking their heads. What had I missed?

"Gerta!" my colleague John Barron called to me as he shouldered through the crowd. "Did you hear about the asteroid impact that killed off the dinosaurs?" he asked, eyes sparkling.

John explained that Luis Alvarez, the Nobel laureate physicist from Berkeley, and his geologist son Walter had found a high concentration of iridium in the KPB red clay layer at a rock sequence in Gubbio, Italy. They believed the iridium (Ir) must have come from an asteroid which collided with the Earth. Iridium is an extremely dense silvery-white rare earth element that can only naturally form in the intense heat of dying or merging stars. Iridium is extremely rare on Earth, and most of what is found in the planet originates from the accumulated space debris that coalesced into the planet when it formed billions of years ago. This iridium is found in natural deposits in the Earth's crust or can be belched to the surface through volcanoes or hydrothermal vents. But iridium is also found in the extraterrestrial rocks that continue to crash into the planet's surface. The Alvarezes must have been studying a rock sequence that revealed iridium within the geologic record precisely at the time of the KPB mass extinction when the last dinosaurs were presumably wiped from Earth.

For the next hour, we huddled around the office radio and listened to various stations as they reported this remarkable story. The excitement in the room was palpable, but—as one would expect from a group of scientists—so was the skepticism. As the full details of the asteroid-impact scenario emerged, the corollary hypotheses painted a dramatic scene: a giant impact-generated dust cloud cutting off sunlight, blocking photosynthesis and plunging Earth into freezing darkness. After the dust cloud settled, wildfires engulfed the planet, roasting plants, dinosaurs, and most other animals, eventually extinguishing up to 75 percent of all life on Earth.

We listened, riveted. It sounded like a Hollywood movie. But not one of the scientists bought it. A giant dust cloud cutting off sunlight and blocking photosynthesis, which is the key to life, was pure fantasy. After all, we knew from the fossil record that most photosynthesizers in

10   THE LAST EXTINCTION

the oceans including algae, diatoms, radiolaria, and benthic forams, had survived the fifth extinction nearly unscathed. If the world had been plunged into darkness, why had they survived this global catastrophe? We knew as well that a planet-engulfing wildfire was also impossible, since such a conflagration would have used up all the atmosphere's oxygen long before it could reach even continent-wide proportions, let alone global. Alvarez's hypothesis was too far-fetched to be believed. I didn't foresee it gaining traction.

And despite my skepticism, I couldn't help but admire the audacity of this idea, which seemed tailor-made for capturing media attention and the public's imagination. An asteroid! For decades, geologists had been trying to solve the puzzle of what caused the fifth extinction by keeping our heads down in the dirt. And this man Alvarez came along and looked to the sky.

That said, it wasn't an *entirely* original line of thinking. Through the ages, scientists have proposed various celestial catastrophes as the culprits behind earthly disasters, going back at least to the Babylonian Talmud when the Flood was ascribed to two falling stars (~200 AD), and in 1696 as caused by Earth's passage through the tail of a comet. In fact, the idea of a giant impact having caused the end-Cretaceous mass extinction first emerged in the 1950s. It was put aside due to a lack of evidence. Alvarez now revived the idea and took it a step further by assuming that the anomalous iridium concentration proved an asteroid impact.

Surely it wouldn't be long before scientists outside geology recognized the obvious: The evidence ostensibly underlying the Alvarez hypothesis was problematic. In particular, the idea that the mere presence of Ir pointed to an asteroid collision seemed obviously suspect. One of the arguments to justify the conclusion that there was something extraordinary about the iridium layer they found was that iridium is relatively rare in the Earth's crust and appears in higher concentrations in asteroids and meteorites. But the concentration of iridium deep inside the Earth is nearly the same as it is in asteroids or meteorites: six hundred parts per billion (ppb) inside Earth and five hundred ppb in extraterrestrial bodies. Alvarez had inferred that the iridium was extraterrestrial, wholly ignoring the more likely explanation that volcanic activity brought magma that was

THE BOMB   11

rich in iridium from below the crust to Earth's surface. Once brought to the surface it could be distributed in giant plumes just like other toxic elements and greenhouse gases. Each of the five major mass extinctions in Earth's history was accompanied by such continental volcanism.

I'd never heard of Luis Alvarez or his son before, and it struck me as odd that a physicist would make so bold a claim outside his specialty. I wondered if this was an instance of the so-called "Nobel Disease," in which Nobel laureates assume a kind of godlike status within the science community that pressures and emboldens them to venture into fields outside their areas of expertise. Was Alvarez's impact scenario intellectual hubris, I wondered, or was there something here? If his new hypothesis could be rigorously tested, it would be fantastic for the field. I could foresee the media attention—and research funds—that would flood toward such dramatic studies. Perhaps, I reasoned, some of that funding might even flow in my direction to help me pursue research into an alternative, and more credible, hypothesis: that the fifth extinction was due to climate change caused by tens of thousands of years of major, sustained volcanic eruptions in what is now northwest India.

This theory, well known as Deccan volcanism in India, had been proposed in 1978 by the paleontologist Dewey McLean from Virginia Polytechnic Institute. Dewey had never been to India, but he was well read on Deccan volcanism. He believed cataclysmic eruptions, which have occurred over an area about three times the size of France, released massive amounts of greenhouse gases—carbon dioxide and sulfur dioxide—into the atmosphere, causing rapid and extreme warming. Those volcanic eruptions also globally distributed Ir and account for the anomalous KPB concentrations.

Dewey's theory was about rapid climate warming sixty-six million years ago, but he was also interested in what warnings this ancient catastrophe held for us today. Long before it was settled science, he predicted Earth was approaching another of these calamitous warming episodes due to fossil fuel burning. Our industry, automobiles, airplane flights, and other polluters—produced by technologies that existed for mere decades—were, he alarmingly pointed out, producing the same climate changes that Earth's natural geologic processes took tens of thousands of years to create

12 THE LAST EXTINCTION

prior to the fifth mass extinction. Deccan volcanism not only helped explain the fifth extinction, Dewey maintained, but should be understood as a cautionary tale: We could be facing another mass extinction soon.

The science community viewed McLean's ideas as credible but lacking evidence. These were still early years of systematic studies of climate change, including the study of tree ring growth and ice cores. Tools like state-of-the-art instruments and satellites were only beginning. But climate warming steadily increased, and scientists were alarmed. In 2023, the International Panel on Climate Change (IPCC) pointed out, they were years away from accepting that "the influence of human activity on climate warming [had] evolved from theory to established fact."

Though few believed that Earth would confront a global warming catastrophe anytime soon, Dewey was building his career around this hypothesis, and it was well-known within the field. I didn't know much about Deccan volcanism at the time, but Dewey's hypothesis was more compatible with my observations than Alvarez's instantaneous asteroid-impact catastrophe. My fossil research, along with many others before me, showed that long-term climate warming accompanied decreased diversity and reduced species populations long before the fifth mass extinction. Considering the day's news, my El Kef samples assumed new significance.

Back in my office, the slides shone like a beacon before me. What clues did they hold? What could they tell me about what the world was like, millions of years ago, when the material and life contained in these samples was on the surface? Did the fossilized animals within experience a swift extinction after a cataclysmic asteroid collision? Or was the story slower and more subtle than that? It was an exciting mystery, and I couldn't wait to collect more evidence, to see where the facts would lead.

2

The SEARCH for TRUTH

EARLY IN OUR RELATIONSHIP, MY HUSBAND AND I HAD LONG arguments about the nature of truth. He faulted me for seeing only black and white, truth and lies, with nothing in between. In his view, there are always shades of gray whether in science or real life. Truth is not absolute, he believed, but relative, depending on the circumstances. The same, of course, goes for lies. Majda was a brilliant mathematician, and he believed that absolute truth can only be found in math theorems and demonstrated in numbers and logic. I couldn't argue with that, but I did eventually convince Majda that evaluating truth in geosciences, even tens or hundreds of millions of years ago, is possible based on evidence. You may never find absolute truth in geosciences or any other sciences, but you can reveal ever darker shades of gray through the accumulation of data that support one another across disciplines and yield the most likely interpretation. In graduate school, my thesis advisor Jim Ingle advised, *Always remember that truth in science is only as good as the data you have at a given time and may change with the discovery of new data calling for reinterpretation.* This has remained my guiding principle in search of scientific truth.

When I was just starting out in my career in science, it never occurred to me that something I considered nearly absolute could be viewed by

14 THE LAST EXTINCTION

others as malleable, political, or subject to interpretations. I had much to learn.

The search for truth begins with a hypothesis, which is an educated guess or hunch created to explain some phenomenon. Scientists then test the hypothesis to evaluate its merits. This is a process that should take time, so that many scientists can scrutinize the claim. In an ordinary scenario, when a hypothesis is scientifically tested and confirmed by research, observations, and facts, it turns into a theory. This, however, was not the case with the impact hypothesis, which swelled into an unstoppable media juggernaut in the weeks and months after the announcement. From the moment it was first published in the June 6, 1980, issue of *Science* magazine, the impact hypothesis captured the public imagination. An asteroid hurtling through space . . . a catastrophic explosion . . . a gigantic dust cloud that plunged Earth into freezing darkness and destroyed the dinosaurs. Who could be immune to such a spectacular story? As Australian journalist Ian Warden later reflected, all it lacked was "some sex and the involvement of the Royal Family and the whole world would be paying attention."[2]

Support for the impact hypothesis at this time was strengthened by a growing interest among astronomers to find and catalog potentially dangerous near-Earth objects. The same month the impact hypothesis was announced, NASA's Advisory Council sponsored a "New Directions Symposium" that discussed "Project Spacewatch" at Woods Hole, Massachusetts, which piggybacked on the impact hypothesis, its publicity, and the participation of Alvarez's group. The ostensive purpose of Spacewatch was to predict and prevent a calamitous Earth impact by blasting from the sky any threatening asteroid, meteorite, or comet.[3] It may not be a coincidence that—threatened by budget cuts from the recently elected president, Ronald Reagan, (who had run on a platform of ending "big government,")—NASA was drawn to scientific research that could both be justified as in the interest of national security and could also obtain, as one NASA official put it, "high scientific yields at low cost." According to a *New York Times Magazine* piece about a symposium on meteorites that was held at New York's Museum of Natural History in 1981, "meteorite research [has] increased dramatically in scope and excitement during the last few years," with NASA's budget

THE SEARCH FOR TRUTH   15

for meteorite studies having grown sixfold in the preceding eight years. This trend would continue throughout the 1980s, with universities and independent scientists pitching in to track the skies for potential planetary dangers and scouring the Earth for large craters. By the time the 1990s arrived, NASA was joined by the US Air Force in devoting significant government resources in efforts to find extraterrestrial objects that might visit destruction on humanity.

The impact hypothesis was good for business, and many seemed to be benefiting from it. The problem was only scant evidence existed to prove the hypothesis was true. No one seemed to care. With growing alarm, I watched as the line between truth and lies, reality and fantasy, blurred.

Never in my nascent career had I witnessed a scientific hypothesis—especially one so weakly supported—translate directly into such expensive programming without ever testing its merits and supporting evidence.

As impact theory received more and more publicity, Luis Alvarez dominated the media coverage while Walter faded into the background. In public appearances, Luis was animated and convincing, and he clearly loved media attention. Nearly seventy years old at this point, he wore glasses, tweed sport-coats, and an easy smile, and projected a patrician geniality. But he was no cuddly grandfather.

The more I learned of him, the more wary and unimpressed I became. There was no denying that Luis Alvarez had led a remarkable career. But for me, his many brilliant discoveries and inventions were overshadowed by the stain of his involvement in the Manhattan Project, where he had served as a senior member of the science team that created the nuclear bombs dropped on Hiroshima and Nagasaki.

His character struck me as even more slippery when I learned of his betrayal of Robert Oppenheimer, his former superior at the Los Alamos laboratory, at the 1954 hearing of the Atomic Energy Commission. The hearing, to determine if Oppenheimer should have his security clearance revoked for un-American activities, was pure McCarthyism, and Alvarez was one of only five physicists willing to participate.[4] Alvarez's readiness to implicate Oppenheimer stemmed from Oppenheimer's opposition to Edward Teller's hydrogen bomb plan, of which Alvarez would be a leader. At the hearing's conclusion, Oppenheimer's security clearance

16   THE LAST EXTINCTION

was revoked, ending his career with the government, and exiling him from his life's work.

Alvarez's career struck me as strange. It veered back and forth between legitimate groundbreaking science and bizarre forays into sensationalized investigations that courted strong public interest. In addition to his work on the Manhattan Project, his early work included the development of innovative radar systems that were used by the military to detect enemy aircraft and submarines, and to safely guide planes into landing.[5] Yet in 1967, Alvarez made headlines for his misadventure using x-rays to scan the Chephren pyramid in Giza, Egypt, in search of secret chambers and treasure, which failed in 1969.[6]

He then won the Nobel Prize in physics in 1968 for his invention of the hydrogen bubble chamber, which enabled the detection of new subatomic particles. A year later he was delving into analysis of photographs of President Kennedy's assassination, supposedly searching for new evidence pertaining to the timing and nature of the gunshots. And now, he was venturing into geosciences, a field in which he was entirely inexperienced. I struggled to make sense of him. Was this a brilliant man whose unbound curiosity propelled him into diverse fields, or an attention-seeking egomaniac unaware of his own ignorance?

What alarmed me most was the way Luis Alvarez and his followers played fast and loose with the truth in the effort to promote his ideas. Geoscientists search for evidence that can reveal the deep past and adjust interpretations accordingly, but Alvarez had no use for evidence that didn't fit his impact hypothesis.

Within weeks that summer, toxic in-fighting took root within my field. The impact hypothesis divided scientists into two distinct camps: the believers and the skeptics. The believers, a group that included renowned theoretical paleontologists David Raup of the University of Chicago and Steven J. Gould of Harvard University, pledged their allegiance to Luis Alvarez. Most paleontologists and geologists, however, remained skeptical.

Skeptics within the geosciences circled the wagons and prepared to take Alvarez to task. Paleontologists were Luis Alvarez's Achilles' heel, because we had a trove of accumulated fossil records that were at odds

THE SEARCH FOR TRUTH   17

with his hypothesis. Dinosaur experts understood that these giants decreased in diversity and abundance over millions of years and began their final decline during the last million years prior to the fifth mass extinction.

But Alvarez dismissed this strong record of fossil evidence and did little to conceal his disdain for paleontologists. In the minds of many scientists, there is an intellectual hierarchy among the disciplines. Mathematicians rank at the top of the intelligence ladder, with physicists just a rung below. As you climb down the disciplines through chemistry, biology, and earth sciences, you will eventually reach the lowly bottom rung, where we paleontologists and geologists pound away at our rocks with hammers to recover clues to ancient environments.

Confident in his superiority as a scientist, Luis Alvarez had no problem publicly attacking and bullying scientists he saw as beneath him. A 1988 *New York Times* interview, for instance, contains a line he was fond of repeating: "I don't like to say bad things about paleontologists, but they're not very good scientists. They're more like stamp collectors."

Every discipline has its own way of finding answers to questions, and in this sense, few fields are further apart than physics and paleontology. In his lectures, papers, and autobiography, Luis Alvarez repeatedly expressed a belief, commonly held among physicists, that two proximate events likely have a causal relationship. This makes sense when one is focused on matter and the universal forces that are acting on it. If you are accustomed to studying the immediate interactions of matter and energy—say what happens after you toss a ball in the air or split an atom—then it may seem reasonable to conclude that the presence of seemingly anomalous concentrations of iridium next to evidence of a mass extinction in the same sediment layer from sixty-six million years ago is not a coincidence. These two events must be related and that must mean that the impact caused the mass extinction.

But two events juxtaposed without further evidence does not prove causality in the geological record because enormous amounts of time are compressed in sediments. In Gubbio, Italy, where the Ir anomaly was first discovered, a millennium is compressed in a few millimeters of clay at the mass extinction boundary. But the Ir peak in this layer is spread

18   THE LAST EXTINCTION

out over tens of centimeters upwards. This pattern, reflecting events from sixty-six million years ago, defies a simple cause-and-effect interpretation. These events—peak iridium concentrations and the mass extinction—are possibly tens of thousands of years apart and not evidence of causality. Earth's environmental history is too complex with many unrelated variables in a highly condensed sedimentary record that does not support causality by simple association.

Alvarez had no use for such a confounding and complicated framework, and neither did the media. As the summer of 1980 stretched on, the impact hypothesis leapt from scientific idea to public obsession. It is easy for a journalist to celebrate an exciting new theory that provides a simple, comprehensible answer to a long-debated mystery. It is much harder for a journalist to master and present the scientific evidence that argued against the simple theory—that there were many different sources of iridium, for instance, or inherent difficulties in dating sediment layers, or planktic foram species. And besides, who wants to ruin a good story?

We paleontologists had plenty to say in rebuttal. The problem was, we couldn't say it—or at least not quickly enough. The science community's method of submitting papers to science journals and waiting for peer reviews could take a year, with additional months added for revisions, if accepted. Meanwhile, breathless reports of the new theory spread rapidly to the public through print media, television, and radio. In the early 1980s, we were still years away from the opening of the internet to the public, so there was no way to contest it.

While opposition papers wound their laborious way through the peer review process, I watched, stunned, as Alvarez's story ran wild. A tsunami of media attention for Alvarez, combined with the near absence of critical articles, sent a clear but false message of consensus. Scientists not on the Alvarez impact bandwagon had effectively been preempted.

Those of us in the opposition bunker, including Dewey McLean and two geologists at Dartmouth College, Charles Officer and Charles L. Drake, commiserated. *How is this happening?* was the question on all our minds. We didn't yet know that Alvarez's loyalists were blocking publication of opposition research through negative peer reviews, but we knew something wasn't right. Alvarez appeared to have a magic hold over the media. *Science*, the most widely read science magazine in the world and one of the most prestigious, kept churning out numerous supporting papers by the impact group along with highly favorable comment pieces by its own journalists.

I've always been wary of doing the popular thing, and the fanatic new interest in KPB studies turned me off. With the iridium anomaly believed to be evidence of the asteroid impact, the battle for proof of the mass extinction rested primarily on species survival rates based on the abundance (or absence) of planktic forams. And yet I sensed that it would be difficult to research this problem in a neutral environment. I was less than one year into my El Kef foram research on the mass extinction and far from reaching any conclusions. But I sensed that any evidence I uncovered would never be accepted if it didn't support the new mass extinction scenario. This meant I would have to jump on the Alvarez bandwagon, agree with whatever scenario he and his followers wanted to spin, and dismiss any evidence that didn't fit. I would have to give up being my own pathfinder and become a follower.

That summer of 1980, I decided to delay my mass extinction studies for five years, expecting that the Alvarez drama would blow over by then and I could proceed without interference to find the real cause of the fifth mass extinction and the death of the last dinosaurs.

I couldn't have been more wrong.

Despite a lack of evidence, it only took two years for the impact hypothesis to catapult to proven theory. You could date this change to April 28, 1982, when Luis Alvarez delivered a hotly anticipated lecture to the National Academy of Sciences.[7] Alvarez began the lecture with his usual bravado, casually dismissing any dissent from his point of view in a few sentences. "I think the first two points—that the asteroid hit, and that the impact triggered the extinction of much of the life in the sea—are no longer debatable points. Nearly everybody now believes them." If the weasel words and repeated references to a consensus that didn't exist weren't exactly scientific, they were an effective rhetorical strategy. And it continued throughout the lecture.

"*Everybody* now agrees that the iridium concentrations we find are consistent with the asteroid impact hypothesis," he claimed, and, "*Almost*

20 THE LAST EXTINCTION

everyone now believes that a ten-kilometer-diameter asteroid hit the Earth sixty-six million years ago and wiped out most of the life in the sea." In fact, there was a healthy amount of skepticism about these claims at the time.

But Alvarez did more than just employ debater's tricks. He also got the science wrong. Understanding that the theory of Deccan volcanism posed the most compelling threat to his theory, he proclaimed a volcanic source for the KPB Ir anomaly was impossible because volcanic eruptions would be very slow and accumulate little if any iridium. To make this point, he erroneously compared the very low Ir concentration in the Earth's crust with extraterrestrial Ir content, while ignoring Ir deep inside the molten Earth surfacing with even higher concentrations than in asteroids, meteorites, and comets. He concluded: "The Ir source could only be from an asteroid" and "the Deccan Traps are almost certainly not the source of the anomalous iridium."

What shocked me most about Alvarez's lecture was how he dismissed facts that didn't agree with his hypothesis, but when paleontological data served his purpose, he used it and freely misrepresented the evidence. His description of the planktic foram mass extinction he'd observed himself during a visit to Gubbio's KPB rock outcrop, for example, was as breathless as it was ridiculous.

"You can see them with a hand lens literally by the thousands, right up to the boundary, at apparently constant intensity, and then, without warning, they are gone, right at the clay layer. It was really a catastrophe. They were suddenly wiped out."

Alvarez painted a scenario that fit his impact theory but was false in every aspect: He couldn't see "forams by the thousands" with a hand lens at 15x to 25x magnification, which allows at best to see the rare largest specimens as white specs, nor could he see "constant intensity," or "extinction without warning."

No paleontologist worth his salt would ever make such claims. When examined under a high-powered microscope, as illustrated in their 1980 report, the same sample Alvarez viewed revealed a dwindling foram assemblage with a single large species among mostly small species approaching the mass extinction. This made it clear that large species were already

THE SEARCH FOR TRUTH 21

endangered or extinct well before the mass extinction. Above it there is a great abundance and diversity of small survivors and newly evolved tiny species characteristic of the first fifty to 100,000 years after the mass extinction.

Alvarez didn't understand, or simply ignored, that the foram assemblages contradicted his hypothesis that the impact hit Earth in the prime of life, when in fact most species were already rare and endangered before the mass extinction. Nevertheless, his misinterpretation of the data was viewed as proof of the impact theory among scientists who didn't know better. The impact bandwagon gathered strength and rolled onwards.

From Luis Alvarez, I observed my first lesson in media manipulation. He orchestrated a two-front war—one controlling media and public opinion and the other eliminating or destroying any scientists who questioned or opposed his hypothesis, particularly Dewey McLean.

He adopted a brilliant strategy of sharing privileged information with favorable top scientists across disciplines prior to publication. This large in-group met at invitation-only workshops funded by NASA to plan research that grew out of the impact hypothesis, shared manuscripts, and coordinated publications.[8] From the beginning, NASA was one of the impact camp's greatest financial backers, and scientists benefited from funding for any related asteroid or meteorite linked to extinction topics. By the time the Alvarezes published their paper in *Science* in 1980, they could already call on a wide consensus of key scientists across disciplines for favorable commentary.

Science magazine drove the media hype from the beginning. From 1980 to 1989, *Science* published sixteen pro-impact articles, which were further promoted by thirteen pro-news commentaries by staff writer Richard Kerr.[9] During this time, no articles were published of any opposition hypothesis, including volcanism. So why the bias? It's perhaps no coincidence that *Science*'s masthead was packed with Alvarez protégées and Berkeley connections. Deputy Editor Philip H. Abelson was one of Luis Alvarez's collaborators on the atomic bomb. His successor, Donald E. Koshland, was another Berkeley product. Alvin Trivelpiece, the executor officer of the American Association for the Advancement of Science,

22 THE LAST EXTINCTION

which owns *Science*, was a retired former director of Oak Ridge National Laboratory, where Alvarez had worked.

Science magazine's prestige and media clout made it imperative for other media outlets to keep up with pro-meteorite impact stories to sell newspapers and magazines. This avalanche of pro-impact publicity created an aura of inevitability surrounding the impact hypothesis. Yet opinion polls among dinosaur experts in 1985 revealed a mere 4 percent believed a comet or meteorite caused the mass extinction.[10]

The earliest articles challenging the impact hypothesis appeared in 1985, five years after the initial announcement. The virtual impossibility of publishing critical articles was obvious to any scientist not on Alvarez's bandwagon. The mantra of "everybody believes" became the rallying cry of Alvarez's followers.

It's a long-running issue in science that those in charge of journals and funding agencies have a vested interest in propping up the hypotheses that they help erect and which make their names. And so it goes that after the rot sets in, these ideas lumber on; they have been called "the walking un-dead" or "zombie science." Editors, reviewers, and funding agencies unwittingly perpetuate zombie science, which differs from simply bad science in that it's immune to evidence. As a result, zombie science becomes entrenched as conventional wisdom.

The idea that continents were forever fixed in place is the most famous example of zombie science in geology. This was the conventional wisdom for centuries until 1912, when the climate expert Alfred Wegener challenged this idea after looking at a map of Earth's continents and noticing that they fit together like pieces in a puzzle. He proposed that the continents were once combined in a giant supercontinent known as Pangea, then slowly drifted into their current positions. For daring to question ingrained belief, Wegener was incessantly ridiculed, ostracized, and lambasted at meetings with his peers. Chastened, Wegener returned to his native Germany and resumed his polar climate research. He died in 1930, at the age of fifty, when his last Greenland expedition ran out of food. He was vindicated in the late 1960s when scientists studying the seafloor around undersea ridges discovered that upwelling magma could move continental plates. This offered irrefutable evidence that the

THE SEARCH FOR TRUTH   23

seafloor was spreading. With this discovery, continental drift morphed into today's Plate Tectonics theory. The tactics of Luis Alvarez and other prominent scientists who have advocated for the impact theory have followed a similar pattern to those who attacked Wegener for contesting the fixed continent theory: Tolerate no opposition, deny all contrary evidence, and ostracize and publicly ridicule opponents.

In step with the rise of the impact theory, a new and disturbing definition of "consensus science" emerged: If a large group of scientists agreed that a particular theory is true, it was true. This philosophy troubled me from the start because it supplanted evidence with popularity. This understanding of "consensus science" still dominates scientific discourse.

Among his supporters, Alvarez stoked a feverish support that bordered on religious fervor. His camp tolerated no questioning or criticism of his ideas. Scientists who found ways to voice their dissent met with swift retribution. Dewey McLean's volcanism hypothesis posed the most severe threat to the Alvarez impact hypothesis, which led Alvarez to unleash a campaign of systematic public humiliation and career sabotage on McLean from which he never recovered.[11]

When I was a graduate student at Stanford in 1977, Dewey McLean visited his former thesis advisor, Bill Evitt, who introduced me to Dewey. He was a tall, lanky man, conservatively dressed in a whiff of military background. He was intensely private and initially struck me as boring. My impression changed when I took him out for coffee at the Stanford café. When I asked him about his research, he transformed before my eyes, relaxing into a wide grin and animated conversation. His energy was infectious, and we spent hours in freewheeling discussion over the likely causes of mass extinctions and Deccan volcanism. McLean hoped to collaborate with me on these ideas.

In 1978, McLean published his most important paper in Science suggesting that "elevated levels of carbon dioxide in the atmosphere caused the global 'greenhouse' warming at the end of the Cretaceous period." He surmised that the relatively high resulting temperatures would have interfered with the reproduction of dinosaurs, eventually bringing about their extinction.[12]

24   THE LAST EXTINCTION

Dewey McLean's idea about climate change was revolutionary at the time, a topic of much conversation. Climate warming caused by greenhouse gas was a concept few scientists recognized at the time. Even further from scientists' minds was the idea that dinosaurs' reproductive failure could be caused by climate warming. I remember Dewey explaining this to me at Stanford in 1977, suggesting that cows could have the same reproductive problems during rapid climate warming. Still, it would take years to recognize that in India's interior seaway thousands of baby dinosaurs never hatched.

And why hadn't they hatched? The answer came years later as I studied dinosaur eggshells, which were extremely thick and hardened, making it impossible for dinosaur babies to break through the shells.

By 1980, three years after McLean published his paper and two years after my conversation with him, the Nobel laureate Luis Alvarez convinced many astrophysicists, physicists, and other scientists that rare iridium concentrations originated from an asteroid impact caused the mass extinction of dinosaurs, as well as 75 percent of other life forms. The story spread like wildfire around the world, but many geoscientists, paleontologists, and dinosaur specialists remained skeptical. Nobody seemed to care about McLean's climate warming, greenhouse gas emissions, and mass death of baby dinosaur hatchlings.

On May 19, 1981, at the mass extinction (KPB) conference in Ottawa, Canada, Luis Alvarez and his son Walter pulled Dewey aside during a coffee break and threatened to sink his career if Dewey publicly opposed the asteroid impact hypothesis. Dewey refused to cancel his presentation. Subsequently, Dewey told me and others: "I felt a moral obligation to continue my volcanism studies. If I was correct that greenhouse warming could set off mass extinctions, and our civilization now faced a possible greenhouse effect, I had a moral responsibility to sound the alarm."

During the Ottawa meeting, Walter Alvarez, playing the good cop to his father's bad cop, drew Dewey aside and pointed to the crowd: "Dewey, count them: Twenty-four are with us. You are all alone. If you continue to oppose us, you will wind up being the most isolated scientist on this planet."[13] Dewey proceeded with his presentation.

THE SEARCH FOR TRUTH   25

The ferocity of the personal attacks that followed has no parallel in geosciences. In a 1988 *New York Times* piece by Malcolm Browne titled "The Debate Over Dinosaur Extinctions Takes an Unusually Rancorous Turn," Alvarez called Dewey a "weak sister," and falsely claimed Dewey was no longer invited to conferences. Coming from a Nobel laureate, the insults were damning. Dewey's department chair turned against him. Fearing for their own reputations, colleagues who had once been friends shunned him.

The situation took a particularly nasty turn when Alvarez dispatched two of his well-known recruits and supporters, Steven J. Gould of Harvard and David Raup of the University of Chicago, to Virginia Polytechnic Institute to urge Dewey's colleagues to deny his promotion and pressure him to leave. Sending such high-level emissaries succeeded. Dewey's department chair denied his promotion and urged him to abandon his volcanism research.[14] Dewey's health suffered from stress and never fully recovered. Alvarez continued to publicly bully and vilify him until his own death from cancer in 1988, the year McLean finally received his long overdue promotion.

Alvarez's strategy effectively shut out those unwilling to jump on the impact bandwagon. Within a few years, his opponents were silenced, and many dropped out of KPB research. Those who remained no longer obtained public funding because by then NSF had been cut to ribbons. NASA and the Lunar Planetary Science Institute (part of NASA) provided generous funding for impactors. *Science* and *Nature* controlled editorial boards for impactors. Impactor also controlled most journals. This left a "silent majority" of scientists frustrated by their inability to get their research published and afraid of public attacks.

One specific group of scientists stood out as an exception: dinosaur specialists. They are a special breed, and few were willing to uncritically subscribe to Alvarez's theory. Until the mid-1980s, they pushed back against impact theory with the abundant scientific evidence that showed the slowly changing environmental conditions that preceded the fifth extinction and the fossil evidence that described a gradual disappearance of the dinosaurs over millions of years.

26 THE LAST EXTINCTION

An informal survey of attendees of a 1985 meeting of the Society of Vertebrate Paleontologists (which include dinosaurs), captures their early antagonism toward the claims of impactors. Conducted by *New York Times* journalist Malcolm Browne, the survey revealed that while nine out of ten paleontologists were willing to accept the possibility of an asteroid hitting Earth around the time of the dinosaur extinction, only 4 percent believed an extraterrestrial impact caused the dinosaur extinction.

At that same meeting, William Clemens, a well-known dinosaur expert at the University of California at Berkeley, reported his discovery of abundant dinosaur fossils in what is now Alaska's North Slope. These dinosaurs, he pointed out, lived in what was a relatively cold environment (for dinosaurs) that experienced an annual absence of sunlight for many months at a time—conditions that were worse than the direst predictions of popular impact scenarios. Clemens suggested, reasonably, that if these dinosaurs could thrive under these conditions, then it is likely they, and others, could have weathered the much shorter darkening and cooling of the Earth proposed in the scenarios offered by impactors. But subtle arguments like these had little chance at capturing the public imagination against Alvarez's scientifically unsupported version of what happened to the dinosaurs. Alvarez seemed more interested in courting attention by saying controversial things and stoking conflict.

His response to Clemens's argument is a telling example of this. Rather than countering Clemens's fossil evidence with scientific arguments, Alvarez retorted that his Berkeley colleague was "inept at interpreting sedimentary rock strata and his criticism can be dismissed on grounds of general incompetence." It was an absurd thing for a physicist to say about an eminent, trained paleontologist, but it got headlines. A *New York Times* article from January 1988 ran his words under the headline "The Debate Over Dinosaur Extinctions Takes an Unusually Rancorous Turn." In that same piece, Alvarez, who would pass away later that year, elaborated on his need to personally attack his fellow scientists: "I can say these things about some of my opponents because this is my last hurrah . . . I don't want to hold these guys up to too much scorn. But they deserve some scorn because they're publishing scientific nonsense."

THE SEARCH FOR TRUTH 27

Considering this, it is not surprising that, along with Clemens, dino-saur specialists like Jack Horner, John Ostrom, Keith J. Rigby, Robert Bakker, and many others withdrew from debates with the impactors. By the mid 1980s, publishing research that questioned the impact theory had become impossible as the impact camp increasingly controlled edi-torial boards of journals. Research funding for skeptics disappeared and the media was uninterested in publishing evidence that didn't agree with the impact theory. Alvarez portrayed skeptics as old-fashioned, outdated, cranky scientists who were incapable of accepting reality. The field now belonged to the impactors.

3

In the CROSSHAIRS

I WAS NOT INVOLVED IN THESE SKIRMISHES OVER THE IMPACT theory in the first year after it captured the public imagination. After I decided to put my studies into the KPB extinction on hold, I turned my attention to an interim project evaluating the Eocene-Oligocene (E-O) transition, another faunal event. For a fossil nut like me, the E-O transition 33.9 million years ago was an exciting time when Earth experienced dramatic climate cooling and the first Antarctic ice buildup. It set the course to the icehouse world that is now melting away.

Today, we know the Eocene epoch lasted from 56 to 33.9 million years ago when most modern fauna developed. It began with rapid, intense global warming linked to volcanism but resulted in no major extinctions. After the warming began, giant prehistoric mammals evolved and walked on land, including the behemoth *Indricotherium*, which looked like a hornless rhinoceros standing five-and-a-half meters tall and weighing up to thirty-three tons. Other prehistoric mammals included cows, camels, sheep, pigs, dogs, horses, and elephants. Cetaceans, the fully aquatic marine mammals such as whales, porpoises, dolphins, and giant penguins, populated the oceans.

The Oligocene epoch was a transition period that lasted from 33.9 to 23 million years ago. In every aspect, from climate change to evolution and extinctions, the Oligocene was unlike any previous epoch. The first

30 THE LAST EXTINCTION

grasslands appeared as land dried up in vast inland seas that separated Europe and Asia. The first ice sheets developed as the sea level dropped and ice remained in Antarctica. At this time, European prehistoric mammal populations of entire families went extinct, including the prehistoric horses Palaeotherium and camels, the Anoplotheriidae. New life evolved, including the first elephants with trunks, early horses, and small mammals like mice, rabbits, squirrels, and beavers, which survive to this day. Time passed, and temperate climate resumed as volcanic eruptions proliferated in the mountains of San Juan, Colorado, and the Absaroka Mountains of Wyoming. Much of the Oligocene world had changed.

The Miocene epoch lasted from 23 to 5.33 million years ago during a time of global climate warming. But polar icecaps persisted in Antarctica, and do so nearly to this day. During the Miocene, sea kelp forests first appeared and became Earth's most productive ecosystem, with whales and pinnipeds feeding on the kelp. Plants, animals, birds, and microfossils, like forams, rapidly diversified and became highly productive ecosystems. There were no significant species extinctions during the Miocene Antarctic deep freeze, although some species adapted to coarse grassy vegetation and fared well. The Miocene was the best time for life in the seas and oceans, climate, and evolution.

The Pliocene epoch marks the time between 5.33 and 2.58 million years ago, and it experienced episodic climate fluctuations, including a warm period in the middle Pliocene. This was the time when Earth transitioned from a warmer climate to the cooler climate of the Pleistocene epoch. On the evolution front, the first direct ancestor of humankind emerged along with the evolution and diversification of new species. The movement of animals across the land bridge of the Isthmus of Panama, joining the Americas, made the Isthmus a major migration route. New animals arrived from Australia and Southeast Asia, including rats and mice. The Pliocene epoch was the last stage of a global cooling trend that led to the Quaternary ice ages.

The Pleistocene epoch is known as the Ice Age, which lasted from 2.6 million years ago to 11,000 years ago. During this time, glaciers repeatedly advanced (glacial periods) and retreated (interglacial periods). The cyclic glaciation was caused by various factors, including climate,

IN THE CROSSHAIRS   31

ocean, and wind currents. Humans appeared, along with new vertebrates and many sea creatures, including giant animals—the Megafauna, like Diprotodon and Megalania. Sea levels dropped over three hundred feet during the Pleistocene cooling.

The Quaternary Period is the current geologic era, beginning 2.6 million years ago and continuing to the present. This was the time of the last ice age, when large herbivores died off from starvation and hunting by humans, between 50,000 and 2,000 years ago. By then, agriculture, the burning and clearing of land for growing food, eventually led to the industrial revolution. Today, rapid fossil fuel burning and increased climate warming have caused extinctions and near extinctions of many species, and our own survival is in danger. We are on the brink of the sixth mass extinction—perhaps our last.

As I began my research into Eocene to Oligocene climate, life and extinctions in the early 1980s, very little was understood about this epoch. I spent days at the deep-sea core laboratory at the Scripps Institution of Oceanography in San Diego, California, sampling middle Eocene to Oligocene cores from the Pacific, Atlantic, and Indian Oceans. In the lab, I freed the forams from the sediments and began to analyze the relative species abundances, which are key to evaluating environmental changes over time.

The concept was simple. Every foram species thrived in the environment to which it was uniquely adapted. These could be tropical, subtropical, temperate, and cool waters, each of which were modified by variations in salinity, oxygen, and temperature. In its most favorable environment, a species reached maximum population abundance. When an environment grew less hospitable, species populations would significantly decrease. If you observe changing foram populations in seawater you can produce excellent climate records via the foram carbonate shells. But there's a twist: Seawater is stratified like a layer cake of different depth and compositions. For example, low diversity and small species lived in surface waters, at high salinity. These environments are agitated by currents and are not hospitable to large populations of planktic forams, but they do record climate change. In deeper, quiet waters there is lower salinity and high species diversity, and larger species thrived in great abundance.

32 THE LAST EXTINCTION

These contrasting conditions revealed the key to the changing climate and environment through the late Eocene and into the Oligocene.

During this time, five major climate cooling phases marked periods of declining temperatures and loss of species worldwide. The early Eocene was the onset of climate cooling and increased loss of species. In the middle Eocene, cooling accelerated with increased species extinctions and reduced survivorship. Near the end of the Eocene, all exotic Eocene survivors had disappeared during further global cooling. What was left were Oligocene survivors of simple small species, low diversity, and life that was adapted to cold climate. This was the transition from the Eocene greenhouse to the Oligocene, the icehouse world.

In 1982, there was renewed interest in this period of climate cooling driven by the discovery of small (one to three millimeters) glass spherules by Billy Glass, at the University of Delaware, and a member of Alvarez's group. Spherules are the byproduct of an asteroid or meteorite impact, caused when the extraterrestrial rock and the earth it strikes are instantly vaporized into a giant mushroom cloud that balloons into the upper atmosphere, condenses, and then rains down to Earth as perfect tiny glass orbs. These spherules were found in deep-sea core drilled off the Caribbean Venezuelan basin that was dated to the late Eocene.

Because they are direct evidence of an impact, the presence of spherules raised questions about this period of changing climate. Did this impact significantly affect life on earth? How did the changes it wrought compare with the KPB mass extinction? Would an impact have triggered sudden species extinctions? Were there long-term environmental effects? Though I was skeptical of the idea that an impact could cause a planetary-wide mass extinction, I was still very interested in the plausible extent of the damage one might cause. I found the implications of Glass's discovery intriguing, and with my new foram data, I sensed an opportunity to break new ground and test the environmental consequences of impacts. I was also relieved that I could dig into these unanswered questions, so closely related to the impact hypothesis, without fear of being attacked for harboring heretical ideas.

For this new investigation, I had to find impact glass spherules in deep-sea cores across the low latitude oceans into which the spherules would have fallen. This was complicated by the limited amount of sampled material that existed. Recall that cores samples used in research are split into two halves, with one half used for study and the other preserved for posterity in an archive. The problem with this protocol is that it is common for research halves to be extensively sampled, leaving little material for future research. This was true for the samples I wanted to study. Thus, the only chance I had to find the impact spherules was in the untouchable, intact core halves of the archive. If I was not allowed to take samples from the archived cores, how would I be able to find spherules? Fortunately, I was able to devise a new method that was simple and non-destructive.

Glass spherules are very good at reflecting light, which is why manufactured spherules are used on road signs and in the white strips on highways that help keep cars driving within their lanes at night. I used this concept to search for the impact spherules by simply directing a strong beam of light onto the surface of the archive core half set under a microscope and moving the core slowly across the path of the beam while watching out for light reflecting glass. This method was a tremendous success. The glistening glass spherules were easy to spot. My adrenaline rushed with the discovery of impact glass spherules in eight cores. Moments like these are what make science so exciting—the chase, the discovery, the evidence, the revealed truth. From a small column of dirt drawn from the ocean floor I had revealed evidence of a brief, violent chapter in the earth's history that happened millions of years ago. I took no meal breaks, and I could barely pull myself away from the microscope to go to the bathroom. I didn't want to miss even a few minutes of this exciting search.

Beyond the discovery of impact spherules, three factors were critical to understanding the nature of the impact, or impacts, that produced them: geography, timing, and the number of impact spherule layers. Determining the geographic distribution was easy. One needed only to compare the concentrations of spherules between known core locations in the middle to low latitude Atlantic, Pacific, and Indian Oceans. Though the precise location of the source crater was unknown, the data told me it was likely in the Atlantic/Caribbean, because samples from

34   THE LAST EXTINCTION

this area in the same period had the highest concentrations of impact spherules, with their frequency decreasing westward across the Pacific and Indian Oceans. I had already determined the timing of the impact as late Eocene from my foram analysis. The number of spherule deposits separated in time revealed at least three and possibly four different but closely spaced age intervals with glass spherules.

I had discovered a comet shower.

I theorized that a big comet broke apart in space forming a series of smaller comets. The very small comets burned up in Earth's atmosphere, but the larger ones crashed on land or into oceans. The impact explosions left evidence in the multiple impact spherule layers across the Atlantic, Pacific, and Indian Oceans.

As I gazed upon the shimmering spherules, I thrilled at the thought that I was staring at the ancient remains of a celestial fireworks display. Comets are dirty snowballs made up of icy snow and rocks that typically range a few kilometers in diameter but may be as large as forty kilometers in diameter. I closed my eyes and imagined them rocketing through the solar system, leaving behind a dusty trail of rocks and ice. When Earth passed through this cascade of comet debris, most of it would have burned up in Earth's atmosphere, creating a magnificent light show against the night sky. Would *Indricotherium* lumbering across prehistoric Siberian and American continents have raised their heads to the stars to observe it? Far better than any sci-fi or fantasy story, tapping the imagination to envision ancient life on Earth is an aspect of geology that never grows old.

Serial comet showers, if large enough, could have been more devastating than a single large meteorite I proposed in 1983. Over the years, impactors and other scientists have picked up the idea of multiple comet showers in the late Eocene of the North Atlantic. By 1999, the possibility of a large crater in the Chesapeake Bay grew more certain and drilling began. The Chesapeake Bay crater is a twenty-five-mile diameter structure (forty kilometers in diameter) buried beneath three hundred to five hundred meters of sediments in the southern part of the bay. The age of the impact dated 35.4 million years.

IN THE CROSSHAIRS 35

A second structure is known as the late Eocene Popigai impact crater of northern Siberia, which is part of Russia. It is sixty-two miles (one hundred kilometers) in diameter and is dated at 35.5 million years ago. The Popigai impact crater area is completely off limits because of diamond mining that supports Russia's economy.

Popigai and Chesapeake impacts likely contributed to the ongoing climate cooling of the late Eocene. But neither late Eocene nor early Oligocene forams revealed signs of a mass extinction. Subtropical species disappeared gradually and went extinct by the end of the Eocene Epoch. Global cooling and the onset of glaciation began in the early Oligocene, and a small group of diverse cold-water species thrived in cold waters and expanded across the oceans.

The Oligocene Epoch lasted from 33.9 to 23.8 million years. During this relatively short interval the first elephants with trunks appeared, along with early horses. But most significantly, cold climate changed the global flora, and many grasses appeared that produced the first vast tracts of grassland that changed life from thereon to the present.

The Oligocene Icehouse World was in full swing.

Little did I know that during the happy months I was quietly working on the forams and impact glass spherules, the Alvarezes were at it again. In May 1982, *Science* published their discovery of a small iridium anomaly located near the recently discovered glass spherule layer in the Caribbean Sea.[15] They interpreted this as part of the dust cloud associated with an extraterrestrial impact. Billy Glass was part of their impact team and published the discovery of the impact spherules. Both papers concluded that a single impact triggered the E-O mass extinction.

I was dismayed the impact thesis was now being applied to the E-O transition. Had I known I might have stayed away from this topic. I knew my work involving impacts and extinctions would be hard not to attract unwelcome attention from the increasingly aggressive and dogmatic adherents of impact theory, should my findings not support their worldview. I had already seen how they treated colleagues who disagreed, and I worried about the unpleasantness—and potential damage to my career—if I got in their way.

36 THE LAST EXTINCTION

But I wasn't sure I'd be able to avoid such a conflict. I already thought the idea that an asteroid strike was the sole cause of death for three-quarters of species on the planet during the Cretaceous–Paleogene extinction was extremely unlikely. To apply the same weakly supported idea to the E-O extinction was an even bigger leap. For one, the E-O extinction was much less severe than the KPB extinction. E-O species extinction affected some marine mammals and large terrestrial mammals but was not nearly at the mass extinction level. If there were a cataclysmic asteroid strike, you might expect it to unleash the kind of widespread and sudden death that was seen during the fifth mass extinction, but the E-O extinction showed limited species loss spread out over at least a million years. It was much harder to imagine this as the result of a sudden cataclysmic event.

My recent discoveries, however, supported the idea that one or more impacts occurred at the time of the E-O transition. This had implications that were both positive and negative to the impact theory. On the one hand, evidence of multiple impacts would delight Alvarez's camp. But my conclusion that they played no role in a resulting E-O mass extinction would put me directly in their crosshairs.

In the spring of 1983, I submitted for publication in *Science*. My results showed multiple impacts and the absence of a mass extinction. The article, "Multiple Microtektite Horizons . . . : No Evidence for Mass Extinctions" (Keller et al., 1983), appeared on July 8, 1983.[16] The response from the scientific community was remarkably positive. For researchers in astrophysics and planetary sciences this was huge. They observed and studied the heavens for clues that could confirm what we see on planet Earth. And now we had the first proof that multiple impacts—a comet shower—could bring about a major catastrophe for life. For Luis Alvarez and NASA it was an equally big deal, because my research suggested mass extinctions occurring over a long time could be explained by a series of impacts. For impactors, the chances of explaining all mass extinctions by either a single giant meteorite, or a series of smaller comets over a short time interval, had just increased exponentially.

What frustrated me most was that the impactors jumped all over my findings to boost their theories—particularly the supposed sixth mass extinction at the E-O boundary. In due course, that hypothetical sixth mass extinction disappeared.

I felt fortunate and proud to have published my research in *Science* and to see it received well. I never imagined that this would be the only research *Science* would publish by me for the next thirty-five years.

My life was about to be upended that summer, in more ways than one. Princeton's Mathematics Department invited Majda out for multiple visits, and in late spring 1983 I visited Princeton at the invitation of the Geology Department's Chair, Bob Phinney. They had been searching for a paleontologist for a couple of years but didn't find the interdisciplinary type they wanted. Phinney invited me to give two talks that integrate different topics beyond the fossil world. The first talk was on my recent work concerning the Eocene comet shower. It was a great success and much talked about. The talk on day two was titled "Why Deep-Sea Sediments Are Incomplete." In this lecture, I argued that deep-sea sediments can be disturbed by ocean currents and are not the pristine time capsules that many scientists have long held them to be. (Keller et al., 1983). At the time, this subject, which my colleagues and I had heatedly debated with prominent marine scientists for three years, was more controversial than the first. This talk was also successful.

Princeton offered me a job on the third day of my visit. Afterward, I went directly to a Geo conference in Dallas and was amazed that just about everyone already knew of my job offer. Several other scientists had interviewed for the same position and my job offer left some sour feelings among them. Whispers circulated that the only reason Princeton would hire me was because of Majda's fame. To be accused of riding Majda's coattail when all my life I'd been staunchly independent and fought hard for every success was the worst insult I could suffer.

Majda and I discussed refusing the job offer. It would have been easy for me to walk away, but I knew how much the position at Princeton meant to him. He sweetened the deal by offering to buy me a house in Princeton that was large enough to accommodate my fourteen pet tortoises. How could I disappoint this crazy, charming man? I hated to leave California, but my future was with Majda. We decided to take a

38 THE LAST EXTINCTION

leave from our Bay Area jobs and try out Princeton for one year; if we liked it, we would stay.

In late summer, I received another invitation. This one sent my head spinning: It was from none other than Luis Alvarez, inviting me to participate in a one-day meeting for a select group of elite scientists at the Lawrence Berkeley Laboratory that fall. By now, I knew that Alvarez viewed my comet shower discovery as key support for the theory that impacts could cause mass extinctions. It didn't matter that I had also concluded—and even put in the title of the article!—that there was no mass extinction at this time. If I attended, I would get a front-row seat to the inner machinations of Alvarez's media circus. On the other hand, my acceptance would surely be misinterpreted by Alvarez and other scientists as support for the impact theory—that is, unless I was bold enough to speak out during the meeting and incur Alvarez's wrath. I had postponed my mass extinction study for five years precisely to avoid this type of situation.

When I began my science career, I made the decision to rein in my impulsive, rebellious nature for a conservative, slow-prodding approach that I believed was necessary in the search for truth. I thought, more than a little naively in retrospect, that this temperament was nearly universal among scientists. Already at this point in my career, this belief had been challenged many times. But witnessing Alvarez's impact theory juggernaut, with its disregard for contrary evidence and aggressive personal denigration of all opposition, created something like a crisis of faith. Only it was my faith in science. Now I felt like I was being put to a test. Would I have the courage to speak up for the real record and incur his wrath? Uncertain whether I should accept this invitation, I asked two people who I knew well and trusted.

Dewey McLean, who knew what it was like to be criticized by the impact camp, urged me to accept. I felt he wanted me to step into his shoes and continue his fight with Alvarez. This was the last thing I wanted. And yet, Dewey's faith in me that I could speak up for reason and truth appealed to my sense of self and pride I'd forged during my young years. Did I dare take up that mantle again by speaking my mind at the Berkeley meeting? If I did, would I endanger the future of my science career?

IN THE CROSSHAIRS 39

Finally, I turned to Majda, who knew me best and wouldn't steer me wrong. He said: "Go for it, attend the meeting, learn what you can. You are strong and don't have to join them."

"So, you would send your beloved into the lion's den?"

Majda's eyes twinkled. "You've been there before, Gerta. Remember Belize?"

I threw my head back and laughed. Majda was referring to my daring cave adventure at the Las Cuevas Research Station in the Maya mountains of Belize, when I was thirty-four years old. My biologist colleague Betty and I ventured into a kilometer-long cave known to inhabit an ocelot, or tiger cat, that fed on wild pigs. Majda was afraid and stayed behind at the Station, unable to dissuade me from going. Betty and I entered the cave excited and unconcerned, walking down the main tunnel and talking up a storm until suddenly a strong musky scent pervaded the air. I spun my flashlight around the cave and stopped short when the light reflected off fresh blood glistening on the ground. With our hearts racing, we turned around and quickly walked back, following the guide string to the entrance.

"You survived that brush with death and lived to tell the tale," Majda said.

We sat quietly for a moment, contemplating. Majda reached forward and covered my hand with his. "Gerta?" he said. "You can hold your own against anyone."

That fall of 1983, I took my seat in a conference room in Berkeley and furtively glanced around. I could see the room was packed with some of the most powerful men in science. I was among nine other scientists at a long table which had a small podium set up for speakers—I was the only speaker. Luis Alvarez sat to my left, and David Raup, a paleontologist who never cared much about fossils, sat to my right. Then there were four astrophysicists who worked on impact theories, followed by Alvarez's disciples: astrophysicists Richard Muller and Piet Hut, and Walter Alvarez's son the geologist, who sat opposite his father.

The most important people, politically, sat in chairs against the wall. This group included Frank Press, president of the National Academy of Sciences, and Gene Shoemaker, the foremost expert of impacts on Earth

40 THE LAST EXTINCTION

and the founder of the new field of planetary geology. (Gene and I had earlier collaborated on my impact spherule research.) Others included editors for *Science, Nature,* and PNAS. It was an intimidating group. I was by far the youngest person in the room, and a relative unknown. I was also the only woman.

"Most of the people here know each other and need no introduction," said Alvarez, "but there is a new member, Gerta Keller, who just made a very important discovery of multiple impacts and comet showers in the late Eocene. She supports our periodicity in mass extinctions." I bristled at this presumptuous falsehood as he went on, "She'll tell us about this discovery."

To understand what was going on in that conference room, it's important to know the cultural context in which the meeting took place. Accounts of global annihilation were very much in the air that fall. On October 30, 1983, an article about Carl Sagan's "nuclear winter" scenario was published in *Parade.* This theory about the consequences of a global nuclear war posited that planet-engulfing fires would be followed by freezing and mass death. It was, in many ways, like the post-impact scenarios imagined by Alvarez and his allies. After Sagan's piece was published, reporters went wild for stories about nuclear apocalypse, and soon the media was filled with nuclear doomsday stories. These doomsday scenarios were woven into an article published in PNAS in January 1984, in which Raup and Sepkoski argued that mass extinctions weren't rare or unpredictable, but that they occurred every 26 million years and were linked to a repeating pattern of asteroid impacts—this was the "periodicity in mass extinctions" that Alvarez claimed I supported, and the reason I had been invited to this gathering of insiders.

As I learned the particulars of the periodicity theory, my stomach flipped. It soon turned out this "periodicity" was an artifact of the statistical method used. It was also based on false assumptions, like the idea that there was a mass extinction during the late Eocene-Oligocene. Nor was there any evidence that mass extinctions occurred at any regular intervals in Earth's history. Nor was there a shred of evidence of periodic impacts on Earth that astrophysicists proposed. Thus far, the only mass extinction to be linked to an impact was the dinosaur mass extinction,

IN THE CROSSHAIRS   41

and that was based solely, and weakly, on the iridium anomaly. There was still no proof of any of this. With growing horror, I realized the impact camp viewed my discovery of multiple late Eocene impacts as a key pillar, propping up their faulty claims that all mass extinctions are related to extraterrestrial forces. This was the real reason behind my invitation to Berkeley—periodicity needed my data, but the evidence was not there.

I realized that the meeting was not to discuss research, but to celebrate a fait accompli. For the men in the room, the debate was over. With other scientists grinning in self-congratulation, Luis Alvarez proclaimed that the fifth mass extinction was instantaneous, coinciding with the Ir anomaly in the KPB clay layer, which proved the impact caused the mass extinction. As the sole paleontologist (a woman) at this meeting, it was up to me to set the record straight or become complicit in the greatest sham in science.

I raised my hand and said something nobody in the room wanted to hear.

"I have not worked on the KPB mass extinction but am quite familiar with the literature. The overwhelming record is of gradual environmental changes leading up to the mass extinction. Therefore, an impact could not be the sole cause."

Luis Alvarez grew visibly annoyed and regarded me with stony silence—then the discussions continued, as if I had been an errant bird that just crashed into the window and briefly interrupted him. The conversation proceeded as before, now discussing Raup and Sepkoski's baroque mechanism to explain the periodic asteroid-assisted mass extinctions, soon to be known as the Nemesis-Death star theory. It was the first time I'd heard this idea, which posited that the gravity of "Nemesis," a conjectured companion star that supposedly orbits our sun, pulls comets into our solar system every twenty-six million years, yielding periodic devastation.

I raised my hand again. "The evidence contradicts this scenario," I said. "There is no impact evidence for the five mass extinctions in Earth's history, and lots of evidence of major volcanism driving climate change. Also, the decrease in species diversity that precedes mass extinctions indicates severe environmental changes." As I presented my argument,

42   THE LAST EXTINCTION

Luis Alvarez became more and more annoyed but continued to ignore my comments. When the time came for me to present my data on multiple impacts, I couldn't help but notice a supreme irony. I, the only skeptic in the room, was also one of the few people presenting actual facts that could be used in the service of their rickety theory. As I passed behind Alvarez's chair to ascend to the podium, he leaned back and muttered: "Gerta, don't bother us with your data, we don't need it, and we don't wanna know it." I felt like I was back on the *Glomar Challenger*.

Despite the hostility in the room, I gave my fifteen-minute lecture emphasizing the comet shower as well as the lack of evidence tying it to the mass extinction. My lecture ended in stony silence. Everyone waited for Alvarez's reaction. Finally, he praised my comet shower evidence, which gained applause. Then he called for the lunch break. I stepped out onto the terrace and stood by the railing, looking out over the beautiful view of the San Francisco Bay. A large tanker passed beneath the Golden Gate Bridge out into the open ocean. It looked as lonely and isolated as I felt. Accepting this invitation had been a big mistake. They had all made up their minds—data be dammed. After a while, David Raup stood beside me and we admired the view for a few minutes together. Raup had a great mind and was one of the few scientists who had taken an interest in my early work and encouraged me early on. Now he turned to me and said:

"Gerta, let me give you some advice: When you are with a group of very important people, don't contradict them; go with the flow."

"But you know they are wrong," I replied. "The data doesn't support their hypotheses."

"You have a lot to learn," he said. "Be quiet and you will go far. There will be other battles you can win. This one you can't." Of course, Raup was right about fighting another day. But I couldn't be quiet.

Coverage of mass extinction periodicity intensified in January 1984 with the PNAS publication by Raup and Sepkoski. That spring, Alvarez's carefully orchestrated media campaign continued. The April 11 edition of *Nature* published four separate articles, some of them authored by the men in that conference room, each offering a different explanation for this unproven periodicity phenomenon. Was it caused by the sun's vertical motion through the galactic plane, by the gravitational effect of

IN THE CROSSHAIRS   43

a mysterious, unknown companion star to our sun named "Nemesis"? It didn't stop there. Pros and cons were also reported for and against periodicities in Earth's magnetic field reversals. All these scenarios were meant to explain periodic events for which there was no evidence.

But none was more popular than the Nemesis idea which, seemingly overnight, transformed from an entertaining speculation into a *theory*. With its intriguing name and sci-fi overtones, it was an instant hit in the popular press. This hypothetical "death star" even spawned a cover story in *Time* magazine and a lengthy autobiographical account by Richard Muller in the *New York Times Magazine* that was later abridged for *Reader's Digest* (and then later expanded for a 1988 book). The media frenzy plunged me into a dark, gloomy mood as I reflected upon the press's outsized role in promoting faulty science.

Despite my clearly stated opposition, the impact theorists continued to use my work to support their "theories" and misrepresent my conclusions. In the fall of 1984, Piet Hut, the Dutch astrophysicist who was also at the Berkeley meeting, invited me to give joint public lectures at Princeton University about the Nemesis theory and comet showers. I accepted. On behalf of Luis Alvarez, Piet invited me to collaborate on a joint paper on Eocene comet showers, which I agreed. Just two days later, I learned that during a public lecture at the Denver Museum, Luis Alvarez announced "most scientists are now onboard with the impact theory, including Gerta Keller, my one-time enemy."[17] Furious, I called Piet Hut and told him the collaboration deal was off. Alvarez withdrew his co-authorship but insisted the comet shower data was critical for this paper.

Fortunately, I wasn't the only one pointing out the flaws in the periodicity theory. Even as sensationalized accounts of a "death star" were being breathlessly reported in the mainstream press, there was, to my surprise and satisfaction, a swift and fierce critical response from scientists. A major issue was their statistical analysis of the fossil evidence that produced their periodic extinction events. Paleontologists found that neither Raup nor Sepkoski had any field experience and so didn't understand the problems of using datasets based on disconnected museum fossil collections or taxonomy described in the literature. The result was

44 THE LAST EXTINCTION

a classic example of the "garbage in, garbage out" concept well known to computer scientists. In the case of the periodicity results, the data itself was biased toward extinctions and therefore gave false extinction peaks. Paleontologists have long been aware of these problems, a major reason why they, as a group, were reluctant to accept the impact theory. (The only paleontologist in the impact camp was a Dutch scientist named Jan Smit, whom we have already met and will hear more about later.) The brilliant Polish biologist Antoni Hoffman ultimately concluded that the periodicity argument advanced by Raup and Sepkoski was "questionable at all levels," and totally unsupported by evidence.

The final *coup de grace* for periodicity in mass extinctions came in the 1987 *Science* article by David Raup's colleague at the University of Chicago, statistician Stephen Stigler (co-written with Melissa J. Wagner), titled "A Substantial Bias in Nonparametric Tests for Periodicity in Geophysical Data." Long before this publication, rumors had circulated that these scholars had advised Raup that the statistics he was using for his periodicity analysis were inappropriate for the data and would result toward periodicity. Raup ignored their advice. With the crescendo of critique over periodicity, Raup's colleagues felt they had to weigh in, or be seen as silently endorsing the periodicity study.

Rumors swirled that when Raup and Sepkoski heard of Stigler and Wagner's paper, they heatedly confronted him in a hallway at the university and the altercation turned physical.

Stigler and Wagner's article snipped periodicity's last thread of credibility. Periodicity in mass extinctions was dead. Nevertheless, Raup clung to the idea of the death star Nemesis until his death in 2015. He wrote in the epilogue of his 1999 book, *The Nemesis Affair: A Story of the Death of Dinosaurs and the Ways of Science*, "The Nemesis theory may turn out to be a major step forward in our understanding of the natural world, or an embarrassing period of near insanity in scholarship." It proved to be the latter.

Despite all these doubts, NASA continued to fund research into Nemesis for years, with Richard Muller at Berkeley and Luis Alvarez leading the charge. In his memoir, David Raup mused that Muller's persistent research may have been driven by dreams of winning the Nobel Prize.

IN THE CROSSHAIRS   45

Nemesis, of course, was never found and faced a quick death in the earth sciences. So did all the other periodicity scenarios that had sprung up like mushrooms during this period of bizarre and unbridled speculations, including the periodicity of impact craters, periodic magnetic reversals, periodic links to volcanism, and comet showers. It could not be otherwise since they were all built upon the false periodicity of mass extinctions.

I got some insight to this question from Princeton's famous physicist Freemen Dyson, who was a vocal supporter of both the impact theory and Nemesis theory. Freeman used to hold weekly meetings at his home in Princeton to which he would invite a wide array of academics and thinkers from different disciplines. At one of these salons I asked Freeman, "Why do you believe so strongly in the Nemesis and impact theories when the evidence does not support it?" I remember Freeman smiled and said, "Because they are beautiful theories. We don't care about evidence, it's immaterial in physics." Startled, I replied: "But science is about the evidence that helps us find truth. If you discard the evidence, you will never find truth." To which he replied, "You will never know the evidence is real or what it means."

Within the space of 1983 to 1987, two distinguished physicists, Freeman Dyson and Luis Alvarez, had each deprecated the importance of scientific evidence to me. It was a preposterous time in my field, when the fantasies of impactors and periodicity advocates turned science upside down. It took a few years for true scientists to shut off this insanity. But within two years, a new impact war started up, one that endures to this day.

4

SNOWBIRD

In late August 1984, Majda and I arrived in Princeton with just one suitcase each to begin our trial year in our new positions. If we liked it, we would stay, otherwise we'd return to the Bay Area to our old jobs.

To my great surprise, Princeton delighted me. I loved the lush, green landscape, the many parks and open spaces, the quaint college town surrounded by farmland. Tree by tree, the fall turned into a rich palette of colors unlike anything I had ever seen before. Every day while walking to the campus we gazed in wonder as the leaves transformed from green to yellow to red and slowly drifted to the ground.

Workwise, Princeton was a mixed bag. I was mostly off to a good start in my department, except for a few memorable setbacks. Al Fischer, my retired predecessor, was supposed to clear out his office by the start of the fall semester but he didn't get around to it for an entire year. In the meantime, I was given a windowless undergrad study room that was not at all suitable for a claustrophobe like me. The cold weather and short days reminded me of my childhood in Switzerland. Within two months, I told Majda I wanted to return to California. My increasing unhappiness was not lost on the faculty and by early December they had voted for my early tenure. Majda begged me to stay through the spring semester

48 THE LAST EXTINCTION

and we spent most of January in the warmth of Cancun, at the tip of Baja California.

Then the seasons changed. Springtime in Princeton is a dream, mixing southern and northern flora at their intersection, creating a palette of exotic and native flowering plants I'd never seen. I realized that if I could adjust to New Jersey winters, I could create the flower gardens of my dreams. My department promised to have my office and lab renovated by the 1985 fall semester. We accepted our positions at Princeton. Majda was happy. Instead of buying a large house with room enough for my fourteen tortoises, he agreed to build a smaller house on a quiet lot with much open land adjacent to the Mountain Lakes Park. Here I had real land at my disposal, to plan my garden and plant endless flowers. In time, I created my childhood memory of the Siberian Iris field in the swampland below the farm where I grew up in Salez, Switzerland, which was forever imprinted on my mind with its blue flowers gently swaying in the wind like ocean waves. In my garden now, every spring, thousands of blue and purple irises shimmer and gently sway in the wind against the forested background. Our East Coast life was underway, and my science realigned itself into its groove.

There's only so long you can stand with your toes at the edge of the diving board before it's time to jump or retreat. By late 1985, I'd waited five long years to resume my KPB mass extinction studies, with my El Kef samples carefully packed away in storage. This was when I decided the hell with it. I'd waited long enough. It was time to seek the truth about the fifth mass extinction, and if that meant a headfirst dive into the deep, so be it.

I resumed my delayed research of the El Kef samples with ferocious passion. My return to this research felt so natural and right, it was as though I was a magnet snapping back into proper alignment. Once more, I was working in pursuit of my dream, and the joy I felt spurred inventive thinking and tremendous energy. First, I analyzed the large forams about three to four millimeters in size. These are the beautiful, ornamented, large species that everyone loves to work with. Delicate and complex, and uniquely adapted to their local environments, there is more species diversity in this group of forams, but also a lower number of individuals per species. They thrived in quiet waters at 200-to-250-meter depths and persisted happily for hundreds of thousands of years with little competition. When climate changed, this specialized group couldn't adapt to new environmental conditions. Only a few species survived up to the KPB, and none are found after the mass extinction.

The middle-sized and smaller species are the workhorses of the foram world. They lived in relatively shallow waters that were agitated by currents and changing climate. Their shells are robust and ecologically tolerant with simple shapes and ornamentations. The shells of these species are made of chambers consisting of two or three rows and they can be grooved, striped, or spirally coiled. One or two of the most robust species dominated the environment and controlled up to 70 percent of the most common species. About a dozen species went extinct at the mass extinction, though a few were able to linger for a time by adapting to the high-stress environments around the mass extinction by "dwarfing," or reducing species size. The sole surviving foram, and the smallest and most primitive type, is *Guembelitria cretacea*. This species continues to thrive and is the sole survivor today.

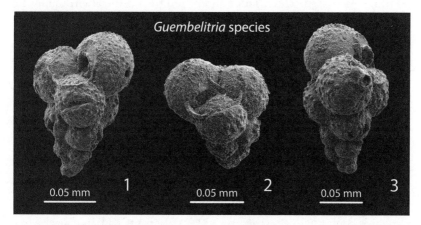

Guembelitria cretacea *is a unique tiny species, the size of a flyspeck, with a shell of three perfectly spherical chambers that increase exponentially in size. This species is also "a disaster opportunist," meaning this foraminifer thrives in high-stress, marginal marine environments. It is the survivor through the ages. During good times when most species thrive,* Guembelitria *recedes to poor low oxygen environments, but thrive again when most species go extinct as during the KPB mass extinction. For full color graphics, visit diversionbooks.com/color-graphics-the-last-extinction.*

50   THE LAST EXTINCTION

Guembelitria cretacea is very small—the size of a flyspeck—with a shell consisting of three perfectly spherical chambers arranged in three stacked rows with exponentially increasing chamber size. It is also, crucially, a "disaster opportunist," meaning this species of foram thrives in high-stress, marginal marine environments. As other species struggle, *Guembelitria* proliferate, choking out other species in a way similar to how red tide blooms create anoxic conditions that suffocate fish and many other organisms. And yet, during good times, *Guembelitria* recedes from normal marine environments, withdrawing to toxic locations along marine margins.

The fifth mass extinction was, for *Guembelitria cretacea*, a very good time indeed. Samples from this period show this tiny foram representing as much as 90 percent of the animal species found in a sample. And this sudden spike seemingly arose out from nowhere. In fact, in these early studies of forams from before the KPB, there was no evidence of *Guembelitria* at all, a fact that mystified scientists at the time. Nevertheless, this sudden spike in the *Guembelitria* population at the time of the mass extinction gave the impression that some catastrophic event had occurred, which suddenly created the hostile environment in which this hardy survivor species prospered.

But I soon found out this assumption was wrong. The mysterious absence of *Guembelitria* before the KPB mass extinction was only a mystery because scientists had not adjusted their microscopes to see just how small these species were before the KPB mass extinction. As soon as I adjusted the microscope, I began to see abundant tiny *Guembelitria* populations before the KPB mass extinction. I also began to see a pattern. As I looked through different sediment layers leading up to the KPB, I saw a gradual increase in species populations, but not the species size.

Below the KPB, the smaller sized species at El Kef and Elles in Tunisia remained common during the first 300,000 years pre-KPB as *Guembelitria* averaged 25 percent in abundance. Between 245,000 and 210,000 years pre-KPB, *Guembelitria* reached 45 percent abundance. By 191,000 to 111,000 years pre-KPB, *Guembelitria* had reached the disaster opportunist level at 60 to 70 percent abundance. Thereafter, abundant disaster opportunists continued to increase up to the KPB mass extinction at greater than 90 percent. But species size doubled and tripled from Tunisia to India (Meghalaya) for a short time interval in the earliest Paleocene.

The evidence strongly suggests this mass extinction was selective and progressive rather than instantaneous. The complex pattern of selective extinctions and evolution mirrored the changing environmental conditions both pre- and post-KPB. Disaster opportunists are the long-term survivors through the ages and therefore could not account for mass extinctions. The message of my study was evident—the mass extinction was neither instantaneous nor long-term.

Although we didn't know it at the time, the reason for the long period of disaster opportunists was Deccan volcanism, which peaked about 225,000 years pre-KPB. Subsequent eruptions continued through the KPB mass extinction. Remarkably, this hardy *Guembelitria* survived through the ages, and today is the morphological equivalent that thrives in highly polluted toxic environments of the coastal areas in the US, India, China, Asia, Europe, and elsewhere in the world. These are today's disaster opportunists, the harbinger of the sixth mass extinction looming soon.

There was something special about my El Kef and Elles studies, and I hoped the breadth and depth of my data could sway the minds of scientists who championed the impact theory. My research had been conducted very carefully and was thorough and unique in its exceptionally high sample resolution analysis. I didn't see how anyone could dismiss the data. But my El Kef results put me at odds with Jan Smit, a Dutch foram specialist, a man I was not eager to make an enemy.

When Smit was a graduate student in 1979, he met Walter and Luis Alvarez, who first proposed the asteroid impact hypothesis. Smit instantly jumped on this idea and later worked as postdoc with Walter Alvarez. I first met Smit in 1983, the day I presented my lecture at UC Berkeley on "Multiple Impacts but no Mass Extinction in the Late Eocene." After my talk, Smit invited me to the paleontology lab and began an awkward conversation about how he met the Alvarezes at a conference in 1979 when they first presented their impact hypothesis.

Smit told me that before he met the Alvarezes he was not sure he wanted to pursue a career in paleontology. Then they told him about

52 THE LAST EXTINCTION

the impact theory. "This was the best thing that ever happened in my life," Smit said with a starry gleam in his eye. "Before the mass extinction nobody was interested, but afterward everybody was interested."

By 1985, the interpretation of the mass extinction presented by Smit was the norm for impactors in Alvarez's camp. Smit's foram mass extinction marked fifty-five species suddenly extinct by impact. But there was no real data.

In 1988, I published my El Kef study and presented my lecture on "Impacts and Mass Extinctions" at the second Lunar and Planetary Institute Conference, known as "Snowbird," because it was held at the Snowbird ski resort in Utah. This was my first time at the NASA-sponsored event filled with impact theorists. I was unknown and virtually without allies in this true-believer country. I would be speaking to a hostile audience—I would present my data on the fifth mass extinction and then deal with the wolves.

I may have suffered from a bit of hubris in believing that my scientific findings could change the minds of my colleagues. Throughout the long journey from Princeton to Snowbird, I felt the thrill and anxiety of traveling with a secret. I couldn't wait to share my spectacular results, but I was also filled with trepidation knowing the histories of those before me who dared to present data that contradicted the impact theory. Yet I was confident that my data were scientifically solid and would speak for themselves.

Before my afternoon talk, I wanted to discuss my results with Smit, naively believing he could be convinced to rethink his instantaneous mass extinction when confronted with the real data. Perhaps, I thought, we could even collaborate to apply the same methods to his Caravaca section to reproduce the El Kef results. My intention was to extend an olive branch and preempt any hard feelings he might have over our differing conclusions. Smit agreed to meet with me over lunch.

We met in the crowded conference restaurant and sat at a small table pressed against the wall. I tried to make friendly small talk, but Smit was silent, chowing down his ham and cheese sandwich. So, I talked about my methods and El Kef results. He asked to see my data graphics, so I pulled them out of my bag, and pushed our plates to the side.

SNOWBIRD 53

Smit's eyes popped and his mouth hung open as he stared at my results. They showed an almost steplike pattern of foram species extinctions beginning at the lower layers, continuing through the KPB mass extinction layer, and finally showing a last group of forams surviving into the early Paleocene. There was no evidence of an instantaneous mass extinction as he claimed based on Caravaca. After a long silence, Smit smirked and said: *At Caravaca all species died out precisely at the iridium layer, which proves the impact theory.*

As gently as I could, I told Smit that I'd carefully read his papers and his PhD thesis and found no Cretaceous foram data. His face paled. He appeared stunned that I went to the trouble to check on his data. "The data is there," he sputtered. "It must be a printing error."

I dropped the subject and instead offered to collaborate with him to compare his findings at Caravaca with my own at El Kef. He stared at me but didn't say a word, shook his head, looked at his watch, and rose, saying, "I've got to go," then turned and walked away.

Smit was clearly mortified. I had called him out on misrepresenting his own data. But I was perhaps the bigger fool. I had revealed everything about my research methods, analyses, and what the results meant in the hope of fostering dialogue that might unveil the real story of the mass extinction. I even offered to collaborate on the Caravaca section, which could help Smit save face. I was convinced he left abruptly to plan his move against me during my twenty-minute talk and rally the impact troops to join his attack. His own data and interpretation on which his reputation in the impact group depended was now in danger of being debunked.

My premonition was right. I didn't intend to provoke a fight at Snowbird II, but that was the outcome. During my frequently interrupted lecture, long lines of scientists snaked up in the middle and on both sides of the auditorium behind the three microphones set up for questions. First in line were Smit and Ed Anders, two of the most aggressive and vocal defenders of the impact theory. Although Smit had little to say to me over lunch, he now led a public assault with thunderous aggressive criticism. For forty-five minutes, pandemonium reigned as Smit and his cohorts took turns gleefully trashing my careful research, while I stood horrified onstage, trying to hide my shaking hands. I was never given

54 THE LAST EXTINCTION

the chance to present my data, no chance to explain the record. They were only interested in destroying this woman who dared to question the impact theory.

Smit immediately called for a blind test of the El Kef samples to determine who was right: him or me. Blind tests in sciences are commonly used to test controversial results produced by high-tech equipment in labs. The test assigns samples to three to four other labs to run analyses for comparison. Blind tests are difficult to apply in paleontology and require intensive field and laboratory research.

Smit continued to fire up his base of dozens of red-faced men, who screamed at me while I tried to present my data. They believed themselves to be the arbiters of truth, while I was a heretic. Their angry bullying was too intense to be caused by only an honest scientific disagreement. They didn't like seeing a woman daring to question their belief in the impact theory. They meant to teach me a lesson to shut up, crawl into a hole, and never dare to speak up again.

I remember time slowing down, as I stood there in a trance watching their contorted faces scream in slow motion. It reminded me of the crowd scenes I'd seen in film depictions of witchcraft trials—just before they burned the accused woman.

In the wake of Snowbird II, I became a pariah, shunned by the impact camp. I continued to give talks at large scientific conferences where impactors were the minority, but was still treated to petty, childish behavior. During social events, impactors often turned their backs to me and closed ranks as I approached. During my talks, they shouted insults and profanities. I noticed other scientists began to avoid me, as if they feared being tainted by association.

One memory from this unpleasant period sticks in my mind. It happened during my public lecture at the Museum of Natural History in New York. While I was speaking, this astrophysicist employee continually interrupted loudly, sometimes screaming at me, for daring to question the impact theory. Riled up and red-faced, he screamed, "You should be stoned and burned at the stake!" At the end of my lecture, sympathetic audience members felt they had to shield me from further hectoring as I walked out of the auditorium.

Despite the uninvited drama during my lectures, my conference audiences swelled to overflow. Frequently, attendees (mostly women) congratulated me for my work and apologized for the rude attacks by men, though none dared to voice public support. Even at Princeton, some old guards remained unwavering impactors. As the younger generation changed and advanced science, the old guard remained entrenched in the impact theory.

Majda was aghast at my treatment and became convinced he needed to protect me. I loved him for it, even though we both knew his arsenal was limited. There wasn't much a mathematician, even a famous one, could do to shield me from the impact armies. Majda pressed me to give up the fight, to preserve both our sanity. He meant well, but there was no way I was backing down.

After the Snowbird meeting, the unprofessional, unscientific treatment I received was not limited to bullying at conferences and public events. It also carried over into damaging, anonymous reviews of the manuscripts I submitted for publication, and negative evaluations of my proposals for research funding. Suddenly, all my professional work was rejected by impact supporters. It quickly became clear that the peer review system doesn't work in such polarized controversies, when access to publication is controlled by an ideological camp. My research was torpedoed again and again by the same avowed impact proponents. I had to fight for every ounce of progress in my field.

Again and again, some editors' decisions were overturned, like Jere Lipps, University of California, Berkeley; (Keller G., 1993. "The Cretaceous Tertiary Boundary Transition in the Antarctic Ocean and Its Global Implications," *Marine Micropaleontology*, vol. 21: p. 1-45) who accepted my Antarctic Ocean KPB data as verified. In 2002, Vincent Courtillot, University of Paris, (editor, EPSL) published a review of my paper by the avid impactor, J. L. Kirschvink at Caltech, who had no knowledge of our KPB but rejected it as worthless.[18] PNAS published our paper in 2004 (PNAS, 2004, March 16, vol. 101, no. 11: p. 3731) "Chicxulub crater predates KPB mass extinction" to huge acclaim and publicity, and it became the most cited paper in our history.

56 THE LAST EXTINCTION

Though it was exhausting, I adopted a warrior mentality. I went into every public appearance knowing it would likely involve a hostile encounter, but I had no intention of quitting. On the contrary, I was just getting started. With my return to mass extinction research, I had reunited passion and purpose and felt driven to discover the truth now more than ever.

The Snowbird II conference was the moment I became fully embroiled in "The Dinosaur Wars." When I took the stage in front of that room full of men who brooked no dissent from their theory, I took a professional risk. But I also knew that I had no choice. I couldn't stand by and watch as the facts and data I was trained to interpret and understand were twisted and ignored to support a simplistic and misleading theory.

Many of my friends didn't understand how I could suffer the constant attacks. *Your peers have left you out in the cold,* they said. Not everyone, I replied, although I knew the list of colleagues willing to stand by me was growing small. I will remain forever grateful to scientists like Dewey McLean, Charles Officer, Charles Drake, Vincent Courtillot, William (Bill) Ward, William (Bill) McDonald, Norman MacLeod, Andrew C Kerr, and David J Archibald, who supported me during this dark period of public harassment. There was also a silent majority who backed me, even if no one dared to speak up.

My friends marveled at my unwillingness to back down. But they, like the men who were trying to shout me down, didn't know where I'd come from. They didn't know that isolation was familiar to me and that I knew how to survive. From my early childhood to my adventures traveling the world as a young adult until my arrival in an academic wilderness, I have never feared confrontation or being alone.

5

The GIRL *Who* DIDN'T KNOW HER PLACE

A POOR SWISS FARM GIRL ISN'T SUPPOSED TO ESCAPE THE Rhine Valley, let alone become a scientist at one of the most elite academic institutions in the world. I'm not supposed to be here, and I mean that in every sense. Not in America, not in my career in science, not even alive. If I'd followed my life's predetermined path, I'd have spent my days as a seamstress in impoverished rural Switzerland. But for as long as I can remember, I've defied expectations to reach higher and attain success I've been told was impossible for someone like me.

Born in 1945, in my mother's ancestral home in Liechtenstein, I was the sixth of my parents' dozen children. Papa was a Swiss farmer and Mama an upper-class debutante who married well below her station in life. Within this pack of Keller children, I was known as the crazy one, a wild child. Serious and observant, I was a loner who would take long walks through the countryside. Overwhelmed with children and the demands of the small farm where we settled in Salez, Switzerland, my parents let me do whatever I wanted, which was unusual for the time. *It's your life*, they told me, *you have to figure it out*. So, I became an inward thinker, relying on my own counsel from an early age. My older sisters, themselves still young children, had their hands full looking after my younger siblings, so I was often free to slip away into nature.

58 THE LAST EXTINCTION

The village of Salez is in the Rhine Valley of northeastern Switzerland, less than one kilometer from the mighty Rhine River, which forms the border with Liechtenstein and Austria. On the Swiss side, towering mountains of the pre-Alps rise steeply to sharp peaks along the narrowing Rhine Valley.[19] Stretching northeast from Salez toward the villages of Rüthi and Oberriet, enormous boulders the size of New York City high-rises stud the grassy and forested hills. The boulders are the tumbled remnants of an enormous rockslide from the mountain tops that occurred a few thousand years ago. Wandering among these rocks as a child, I imagined living in a fairy-tale land of giants. Perhaps this is where I got my early interest in geology—staring at these mysterious boulders and wondering what stories they could tell. Every day I roamed the valley to the foot of the high mountains and often wore myself out by charging directly up the mountain paths until I was breathless, my lungs burning and my heart thundering in my ears. No territory was off limits; my only limitation was the distance my legs could cover in a day. I never tired of being alone.

Only after I moved to the big city of Zürich, and later as I traveled around the world, did I realize how lucky I was to grow up in an unspoiled world free of automobile traffic and the polluting smokestacks of industry. Mine was a world where the humming of bees, the singing of birds, the call of the cuckoo, the moo of cows in the fields, and the occasional alphorn echoed through the valley.

I admired wildflowers and picked bunches to adorn every room of the farmhouse. I knew the fields of blue Iris, the meadows of Trollius flowers (Bubenrollen), the patches of violets at the edge of the forest, the liverworts in the mossy forest, and the "Bedseichele," a beautiful small white anemone said to make anyone who picked them pee in the bed. (It was not true. I picked them and never peed in bed.) I knew where the first snow flowers burned through the snow, which meadows boasted the largest primroses and blue bells. They were my secret places, cherished, protected, and guarded. That first farm love, the love of flowers, working in the dirt, the wriggling worms, the smell of freshly turned earth, the creation of habitats for special plants, has remained with me. It's in my soul, my bones, my hands. It gives me strength and inner peace. It

THE GIRL WHO DIDN'T KNOW HER PLACE 59

connects me to the land, keeps me rooted in the reality of life, the cycle of seasons. I'm a farmer at heart.

One late spring day during my childhood, when I was on one of my rambling walks, I paused at a large house on the outskirts of the village. Pressing my face between the rods of the iron fence surrounding the garden, I watched an elderly woman pull weeds from her flower beds. The beds were a riot of color, filled with exotic red, yellow, and purple cup-shaped flowers I didn't recognize, laced by a border of sky-blue forget-me-nots. Something about the woman intrigued me. She was tall and thin, regal somehow, even in her gardening clothes. She was different from the old women I saw in the village, who were usually short, round, and bent over with a crooked back, their faces deeply grooved with misery.

The old woman glanced up and caught my gaze. I was turning to run when she smiled at me and waved me into her garden. I could not resist the urge to inspect the unknown flowers and slowly entered. I pointed to the flowers I had never seen before.

"These are tulips, from Holland, a faraway country by the ocean," the woman explained.

I reached out to touch one of the warm, waxy petals.

"In Holland, they grow tulips in vast fields, like we grow potatoes," she said. "I've visited Holland. Would you like to see the pictures?"

I nodded yes. I instinctively trusted this woman. I liked her kind old face and soothing voice, but I especially liked that she didn't ask me who I belonged to or why I was far from home. As we slowly climbed the stairs to her third-floor apartment under the roof, she introduced herself as Frau Babaly, the Sunday school teacher of the village church.[20]

Frau Babaly sat me down on the couch in her small living room and she brought out a glass of milk and a cookie. Then she placed a stereoscope and a shoebox full of pictures in front of me. Cautiously, I leaned forward to peer into the scope. She inserted a picture and suddenly I was staring into a three-dimensional world of golden sunlight and azure waters. The scene was of a sandy beach filled with people in bright-colored bathing suits. I was in awe. I'd never seen a beach, let alone a bathing suit. Questions bubbled from my lips.

60   THE LAST EXTINCTION

"This is *das meer*," Frau Babaly explained.

One by one she inserted pictures from her exotic travels into the stereoscope and told me stories of each.

I began to visit Frau Babaly daily. I never grew tired of looking through the lens at the pictures, listening to the soft murmur of her voice as the images sprang to life in my imagination. Sometimes she read to me from books. For the first time, I glimpsed the wide world beyond our village: the images of people of different colors, wide streets with cars, buildings the size and shape of which I'd never imagined possible. Frau Babaly took a keen interest in my observations. She seemed to see me in a way no one else did. With her, I could never ask too many questions.

One day several weeks into the summer, I climbed the steps and knocked on Frau Babaly's door, but she did not answer. I waited on the steps in the sun for some time, then gave up and went home. The next day, I arrived again at the usual time, but again she did not answer. For several days, this pattern repeated. Frau Babaly had disappeared.

I didn't know it at the time, but Frau Babaly had become ill and was taken away to a hospital to recuperate. Although I never saw her again, I've carried her memory with me like a treasure. That summer changed the course of my life. In Frau Babaly, I had found a kindred spirit, one who shared and nurtured my intense curiosity of all things in nature and the world we live in. But most of all, during those short summer weeks, my otherwise ordinary afternoons became transformative experiences that awakened in me a growing awareness of this bigger world beyond my village and the Rhine Valley.

My golden years of freedom wandering the valley ended abruptly when I turned six years old and my family's wretched undertaking to tame me into a proper Swiss woman began. Mama was a hopeless house-keeper, so the responsibilities of maintaining the home fell to me and my sisters. When I wasn't washing and cleaning, I worked in the fields. My sister Rosie, a curly brown-haired beauty with an open, friendly face and self-deprecating sense of humor, often walked beside me, trying to catch my eye and cheer me up. It was hard to be cross around Rosie, who had an effortless way of finding the good in any situation, but I still yearned to be far away on my outdoor adventures again.

THE GIRL WHO DIDN'T KNOW HER PLACE   61

With bitterness, I noted that my brothers had far more free time than my sisters and me. They were not expected to do women's chores like keeping the house. I envied the boys and grew to resent them—all but Freddy, who never failed to make me laugh. The second-oldest boy, Freddy was a character with thick, oily hair, slicked back to look like Elvis. He drove a Harley-Davidson motorcycle and played the trumpet. Because Mama could not stand listening to Freddy practice, she sent him up onto the roof of the barn to play. The sound carried throughout the valley, irritating our neighbors, but not even they could remain angry at the charming Freddy.

My long-awaited first school year began after my seventh birthday. I became addicted to books and devoured the entire first and second grade school library before the end of my first year. One particular book about the celebrated doctor Albert Schweitzer in Africa stirred my interest in medicine and helping the poor and sick. I hungered for stories about the world beyond Switzerland. Every page I turned confirmed my certainty that a different life awaited me, one free of the drudgeries of washing and cleaning and weeding the fields, where girls could be as daring as boys.

By the age of eleven, Swiss school kids must make a choice about what profession they will choose for their lives. I declared that I was going to be a doctor like Albert Schweitzer and help the poor in Africa. My aspirations raised alarm, and a week later I was interviewed by a school psychologist sent by the state. Her goal was to keep students on their socially predetermined paths in life, informing me that my career options were maid, salesgirl, or seamstress. I appeased her by agreeing to the last of these options, telling myself I was interested in fashion. This was trying enough, but the most painful pressure came next.

A few days later, a priest arrived on the farm. Mama welcomed him in, and the two spoke in private in the house. When the priest emerged, he invited me to walk with him. All my senses were on high alert as we began an awkward stroll along the Wisla. He asked me many questions about my life, interests, and aspirations. I never mentioned religion. At the end, he turned to me and solemnly asked if I had a calling to serve God. I said no.

When the priest and I returned from our walk, we found Mama in the kitchen rolling out dough. "It's too early to take her into the nunnery,"

62   THE LAST EXTINCTION

the priest said. "She hasn't had the calling." Stunned, I looked to Mama, expecting her to protest that there had been some misunderstanding. All I saw was her disappointed expression. With horror, I realized that my parents had planned to send me to the nunnery without consulting me. I stared into my mother's face, but she wouldn't look back at me. She busied herself wiping the flour off her hands onto her apron as she prepared to see the priest out. *How could you give me up?* I asked her silently.

Neither Mama nor I ever spoke of the incident again. Something inside me gave way, and I stopped trying to understand her. Our relationship remained civil, but we would never again be close.

Freddy returned home from Zürich the following weekend. When he heard of my intended sacrifice to the nunnery, he took me for a ride on his Harley-Davidson. We rode on top of the Rhine River dam, then took the access path down to the riverbed. The water level was low. Rocks, boulders, and tree trunks had accumulated over the storm season and created a large beach-like area over half the width of the Rhine. Freddy wandered out toward a giant tree trunk, and I followed. Together, we sat on the trunk and listened to the gentle rumble of the water lurching around the debris.

Like me, Freddy wanted to be more than what his teachers told him he could be. After being told he should be a farm worker, he had gone to Zürich, where he took a well-paying but dangerous job switching rail cars at the train station. But his real aspiration was to be a policeman, and he was saving money to pursue that dream. After I told him about the visit from the priest, and my anger at the family for abandoning me, Freddy put his arm around my shoulders and encouraged me to leave the Valley as soon as I could. "You are not a black sheep," he told me. "You're smart and you know what you want to do in life. Stay on your path."

When it was time to go, Freddy drove the long way back through the Rhine Valley to Salez. By the time we reached the farm, my heart was light and happy.

In June of 1958, just two months before his twenty-third birthday, Freddy was killed by a fast-approaching train on the wrong track while working on his job at the Zürich train station. Through the night of his

THE GIRL WHO DIDN'T KNOW HER PLACE 63

death, our parents wailed in pain. The news numbed me. I couldn't cry. I was stuck in a world of disbelief, waiting for a call that it was all a mistake.

After Freddy's death, my view of life soured. Swiss life for girls seemed so restrictive, even sinister, confining us to predetermined paths. I still had one year before I was to embark on my future career as dressmaker turned fashion designer. In the fourth grade, I was considered intelligent enough to attend Realschule, a separate school system for good students. Through Realschule, students could pursue apprenticeships that might lead to an actual career. Of course, the definition of career, and of education for that matter, was vastly different for girls and boys. At Realschule, the boys learned math, physics, and chemistry; girls learned housekeeping, cooking, and cleaning. Needless to say, I hated it.

At the age of fourteen, at the end of my second year of Realschule, I dropped out to fast-track my two-and-a-half-year dressmaking apprenticeship. Weekends I spent with Rosie working at the restaurant in Arbon, fifty kilometers from Salez, where she waitressed long hours from 7 a.m. to midnight, with a two-hour break. I helped her clean tables, wash dishes, and took over her job waitressing during her break. By midnight when the restaurant closed and she had nudged the last customer out the door, I helped her clean up, scrubbing the tables and the floor to have everything ready for opening time. For all this hard work, her pay was just room and meals, and maybe a 10 percent tip, if customers were willing.

At the end of the weekend, Rosie always gave me half of her tips, an overly generous amount to add to my savings to pay for fashion design school, which I planned to attend after my apprenticeship. Those years with Rosie were a happy time for me and relieved me from the drudgery of the seamstress life. Three years later, when that savings had disappeared because Mama needed the money, I could not tell Rosie—she would have been furious with Mama. From then on, my life spiraled into depressions; I stayed away from Rosie, afraid to tell her, knowing that she would have sacrificed herself for me.

At seventeen, I left Salez seeking freedom, but instead found drudgery, working long hours six days a week for exploitive pay. With my savings for fashion design school gone, I moved to Zürich and landed a job at DIOR Creations Zürich, the famous Paris fashion name and the top design

64 THE LAST EXTINCTION

house in the city. For months, I slaved away over my sewing machine creating the most expensive dresses for the city's elite upper class, while earning starvation wages. My earnings barely paid my room plus daily soup and bread. Because I had no money left over for the trolley, I walked the three kilometers to and from work every day. The streets were a gauntlet. Men driving fancy cars pulled up alongside me, asking for sexual favors. As soon as I got rid of one car, another would drive up.

I began to lose hope. I had done what everyone asked of me: I'd earned the highest marks, pursued the best career path available, and had even left the Rhine Valley for the big city, as Freddy had advised. Yet, I had only traded one cage for another. I still had no options for education, no way to break out of the straitjacket of poverty. Dior had been my last prospect for a career, but I saw now that I had no future there. No matter how perfect my sewing was or how exquisite the gowns that I created were, there was no upward path for me to follow. I could never save enough to better my situation. In truth, I was barely surviving. I felt myself on the verge of a breakdown.

Rage consumed me on the inside, while on the outside I began wasting away. My meager meals of soup and bread weren't enough to sustain me. I lost weight and began contemplating suicide. For me, a life without education, a life in which I could not be authentic to my true self and free to pursue my dreams, was not worth living. Just barely eighteen, I wanted to get it over with, go to sleep, and never wake up again.

One weekend, I received an invitation from my sister Ruth to stay in the room she kept in Zürich while she was out of town. Ruth and I had never been close, and her invitation was so unexpected I accepted without thinking. Snooping around Ruth's bathroom cabinet, I found a nearly full bottle of sleeping pills. I swallowed them all and woke up three days later, thinking *There must be a better way to die.*

It's a strange thing for a young person to harbor a death wish. I'd made the decision to end my life, yet almost instantly afterward I felt a sense of freedom and clarity I hadn't experienced since I was a little girl, when the world was still vast and beautiful and alive with possibilities. I no longer felt the pressure to conform to what Swiss society expected of me. I didn't need to attain a career that would sustain me into old age.

THE GIRL WHO DIDN'T KNOW HER PLACE 65

For the first time in a very long while, I thought about what I wanted most.

My thoughts drifted back across the years to Frau Babaly and the marvels I glimpsed through her stereoscope on those hot summer afternoons. I had yet to see *das meer*, meet exotic people from different cultures, and explore the wonders of the world. My final goal was to do these things before I ended my life. I quit my job at DIOR and took up waitressing to save enough money to put my final plan in motion.

With nothing left to lose, I vowed to die by the age of twenty-three, the age Freddy died in the train accident. But I promised myself that I would have many adventures traveling the world beforehand. Strange as it seems, by embracing death, I gave myself permission to live, if only for a time. I would be bold and reckless, take risks, and hold myself back from nothing.

In the end, it was easy to leave Switzerland. When I called Mama from the Zürich train station to tell her about my imminent departure to hitchhike around the Mediterranean, she screamed into the phone, "Why can't you be normal and settle down like everyone else!" Then she hung up. I stood at the pay phone and listened to the dial tone for a long time. With my new plan to die by the age of twenty-three, I assumed I'd never hear my mother's voice again. I stepped into the future with the distinct impression I had nothing to lose.

For four tumultuous years, from 1964–1968, I fulfilled my dream of traveling the world. These were my years of living dangerously, so dangerously that staring into the face of death was an all-too-common occurrence. My journey began in Spain, then hitchhiking across North Africa from Morocco through Egypt. I lived on the $1,750 I'd saved from waitressing, spending on average of one dollar a day by staying at youth hostels, eating every other day, and hitchhiking. I would travel as far as I could go on my savings and then return to Switzerland to waitress for three months during the Carnival and ski seasons from January through March (the most lucrative tipping times) and then venture abroad again.

From Egypt, I journeyed across the Middle East, then on to Greece and Eastern Europe. In Genoa, Italy, I boarded a boat to Australia, then

66 THE LAST EXTINCTION

continued through war-torn Southeast Asia to America. As Frau Babaly had promised me, the wider world was more expansive, wondrous, and intoxicating than I'd imagined. Each new culture I encountered was a doorway I stepped through, with an exponential number of doors on the other side, each beckoning me forward, away from the restrictions and constraints of my youth.

I pushed myself as never before. As I crossed countries and continents, the euphoria of my freedom was bound in lockstep with the urgent knowledge that my time was running out. This awareness pressed upon me the way the air feels electric before a storm. It made me jittery, restless, reckless. I was a girl living in a hurry to experience life to the fullest in the short time I had. Even as I reveled in my travels, I remained determined to end my life at twenty-three. My real-life goal—an education—remained out of reach.

At Morocco's ancient Marrakech Plaza, I sat in a circle of old men drinking sweet tea from tiny glasses, while just footsteps away a dentist pulled a man's rotten teeth from his infected gums with a string. In Fes, Morocco's ancient city, I visited the eleventh-century working tanneries and watched near naked men wade in toxic pits to treat the hides in a pervading stink that made me feverish and ill. In Algeria, I drove across a ghostly coastline where the jungle crept toward the azure sea, reclaiming the war ruins of villages, and met child survivors with red-ribbon scars across their throats once cut by French Legionnaires. I met the French commander in chief in the Algerian/Tunisia border town during the overthrow of President Ben Bella's government. I experienced the famed Fata Morgana in Tunisia, lifting my eyes from shifting sand dunes to the sudden joyful sight of an oasis of green palms and a white minaret. For hours as I drove, the oasis remained fixed at a constant distance until it disappeared. I bonded with women soldiers on the Golan Heights and wished I could have been one of them. In Cambodia, I toured silent, immaculate villages that shook with explosions at night, turning the sky a blinding white, then golden yellow and orange.

It was in Sydney, Australia, with just six more months to live until age twenty-three, that I was shot through my lungs in a bank robbery

THE GIRL WHO DIDN'T KNOW HER PLACE 67

and miraculously survived. Later, I met an American student named Tim Callahan, who would accompany me through my war travels into Asia and on to America, where I eventually found my education. As a former marine turned left wing social democrat, Tim was almost as good at seeking out trouble as I was. A student at San Francisco State University, Tim was visiting Southeast Asia to investigate hot spots of US military involvement. Together, we made our way across Indonesia, Thailand, Cambodia, and Japan, meddling in the politics of war and the overthrow of governments, which occasionally put us in the crosshairs of the secret service.

In late January 1968, we landed in San Francisco. I had five more weeks to go before my twenty-third birthday. America was to be a short stopover en route to Central and South America, where I planned to end my life. My plans, however, would soon change radically.

Tim and I rented a room in a three-story apartment building on Duboce Street, near Haight-Ashbury. Around us, the Flower Children movement was in full swing. Each apartment in our building was shared by four or more students, and these students thrived on two things: LSD and conversation. I never tried drugs because I always wanted a clear head and full control of my senses, but I enjoyed listening to the students' impassioned dialogue. They were completely swept up in their ideals; school to them was merely the community through which they advanced the anti-war and civil rights movements. When I confided in some of them that I'd fled Switzerland because I had no opportunity for education, they were incredulous. *Anyone can get an education here,* they said. *Just apply to San Francisco City College.* When I told them I had no money, they waved off my concern. *City College offers financial aid. It's almost free if you're a high achiever.*

I was skeptical, but Tim and the students pushed me to explore it. I sought out an application, and my heart fell when I saw the requirement of a high school diploma. I had no such document, and no experience remotely the equivalent. The thought of trying to explain Realschule and my dressmaking apprenticeship to an admissions officer was exhausting, even laughable. Yet Freddy's words echoed in the back of my mind, *You can do anything you set your mind to.*

68   THE LAST EXTINCTION

The next day, I returned to City College with a completed application. Beneath the high school diploma requirement, I'd checked a box stating that I had no diploma because my high school had burned down. I was given the placement exam and passed.

I was twenty-three years old, the age I'd determined to die, but instead I found a path to the education I'd always wanted and a future worth living. How can I describe the exhilaration I felt my first years in school? I had the overwhelming sense my life had realigned according to its true purpose. Somehow, in spite of all the misery and hardships, and my twisting chaotic travels, I'd arrived at the place I was always supposed to be. The students had been right; I earned nearly perfect scores at City College, and in return they provided me with enough aid that I no longer had to waitress to make ends meet. After two invigorating years that flew by like a dream, I transferred to San Francisco State University and took my first course in geology.

I'd always wanted to study science, but in Switzerland, science was only for boys. Science intimidated me. For two years, I'd skirted around it, taking anthropology and philosophy courses, which seemed the next best thing. Geology seemed an attractive entry point, incorporating the study of spectacular mountains, flora, and oceans rich in fossils. My farm girl love of working in the dirt remained with me. A course titled *Man and the Ice Age*, taught by the Swiss Professor Hans Thalmann, piqued my interest. In his class, I encountered Norman Newell's 1963 seminal paper, "Revolutions in the History of Life." This paper set my mind on fire. Five times in the history of our planet, more than 75 percent of life was obliterated in what Newell called mass extinctions. Each time vast ecosystems teeming with life disappeared. Then somehow, miraculously, life rebooted. *Why? How?* The mass extinctions struck me as a stunning scientific mystery. *What caused a mass extinction?* I wanted to know, *And, if it happened five times before, could it happen again?* If so, this calamity would potentially threaten mankind. I was determined to discover the truth about the peculiar conditions that created and extinguished life. I dreamed of being the one to crack the case of this massive murder mystery.

I asked Professor Thalmann, "Why did you choose geology as your career?"

THE GIRL WHO DIDN'T KNOW HER PLACE   69

With sparkling eyes and a smile, he replied, "If you love to travel, love to sit at the beach and watch the sunset, you should choose geology. There are rocks everywhere, and you can always dream up a proposal to study them and someone will fund your trips."

He then told me he made a life doing just that and enjoyed every minute of traveling the world to the most beautiful and exotic places. Everywhere, he collected rocks to study microscopic foraminifera fossils and pioneered their study to help oil companies find oil reservoirs. Foraminifera are everywhere in the world, he advised, and they have very useful applications in climate research, toxic environments, and all kind of environmental changes. I didn't know what foraminifera were at the time, but if geology could provide a career in research into mass extinctions and combine it with world travel, I was sold.

Tim Callahan and I parted ways in 1973. We had grown apart as I ferociously, almost fanatically, spent my time on mass extinction research. I couldn't get enough of it, and all else took a back seat. I was eager to be on my own, planning my future. Shortly before graduating from SF State in 1974, I won a prestigious Danforth Fellowship that covered all my expenses for graduate school. I chose Stanford University and became one of the first three women accepted into the geology program. I could scarcely contain my excitement. I felt I'd made it and could now look ahead to a life rich in intellectual stimulation and thrilling research.

On the first day of the PhD program, we three women walked across the old quad toward the corner that the geology department occupied, where we would share an office on the third floor. Inside the entry hall, half a dozen professors stood and watched us approach. I smiled, thinking they were waiting to welcome us, but as we passed them to climb the stairs, one of them quipped, "I might say, we did pretty well with our first women graduate students," and everyone laughed.

My face flushed red with fury, and the other women and I exchanged looks of dismay: Sexism had pursued us into the elite Stanford campus. I felt foolish for believing Stanford would be different. It seemed no matter how far I climbed, some men would only see women as a commodity. Shaken, I continued to ascend to the third floor, determined not to let a few insensitive comments derail my dream.

70 THE LAST EXTINCTION

With the fellowship paying all my expenses, I was free to pursue whatever topic I chose, but mass extinction studies were not a topic of research at Stanford or most other universities at the time. For my doctorate research, I chose to study planktic foraminifera—the micro-fossils whose study Hans Thalmann pioneered and turned into a very successful career.

My workdays generally spanned sixteen to eighteen hours, including course work and microscope research of the forams. In my second year of graduate studies, I inherited a large office from a former student and acquired a narrow cot to sleep on. During the long hours working on the microscope, I practiced taking short naps. Thomas Edison was said to have existed on short naps, and I followed his example. Eventually, I trained my brain to fall asleep immediately and wake up at the exact minute I set my mental clock. It worked like magic, allowing me to continue my research into the early morning hours.

Far from resenting the long days, I thrived on them. The research was so much fun: I felt like a detective searching for clues of long-ago crime scenes. Every day I learned more, and uncovered another piece of the puzzle, but still the missing parts beckoned. My mind churned, incessantly searching for that key piece of evidence. There is no doubt I was obsessed with my research, seeking the truth wherever it would lead. I disappointed my fellow students by turning down their invitations to social gatherings. Often, I'd refuse to even take a coffee break. My excuse was that I had too much learning to catch up on, having started my education so much later. Soon they all gave up on me.

I was becoming a loner once again, but my independence had its perks. This was a golden period for me, a chance to feed my curious mind and direct my life exactly as I wanted. It was during this time that I discovered my passion for running. Every day at lunch I ran five miles in the Stanford hills, a vast, beautiful area of unspoiled golden-brown grassy hills and oak trees, high enough to see the skyline of San Francisco in the distance. I became hooked on the runner's high, the brain's release of endorphins that creates a sense of euphoria so intoxicating it transforms running into a feeling of effortless flying. As in my youthful sprints up the winding Swiss mountain paths, I charged the California hills with glee, sometimes feeling guilty as I passed other runners who labored uphill step-by-step. With my feet rhythmically striking the ground, my heart beating wildly in my chest, and air plunging deep into my lungs with every breath, I felt gloriously alive and free.

6

The CRATER

THE YEARS I HAD SPENT TRAVELING ALL OVER THE WORLD IN my youth helped prepare me for my trips to the field. Geology fieldwork will take you to far-flung, exotic places. But not all field trips are successful, and not all adventures are fun. Some can be outright dangerous. In late 1993, an ill-timed research trip to Chiapas, Mexico, dropped me in the early stages of the Zapatista uprising. Images from that trip haunt me still.

Two colleagues and I were driving north from the town of San Cristobal de las Casas into the country through empty streets and empty villages. Freshly dug graves, sometimes dozens of them covered with flowers, crowded every village cemetery. We were silent, afraid of what was to come. Days earlier, when I arrived at the Mexico City airport to transfer to my flight to Tuxtla Gutierrez, someone thrust a leaflet into my hands. It was a warning of an armed conflict in Chiapas, urging all travelers to stay away. How could I have been so stupid to press on?

Arriving at Ocosingo, the largest village, we confronted a grotesque, surreal spectacle. Over a hundred native Chiapas men had been corralled into the village plaza and fenced in with barbed wire. Surrounding the fence, women and children cried out for their husbands, fathers, sons, and brothers, while menacing soldiers with rifles patrolled the rooftops of the houses surrounding the plaza and marched through the streets.

74   THE LAST EXTINCTION

We drove quickly through the crowd, not wanting to stir suspicion and be mistaken for foreign press reporters. Soldiers eyed us warily, pointing their guns at our Volkswagen bus. My German colleague Wolfgang, a paleontologist at the University of Nuevo Leon in Linares, gave them his sunniest smile and waved, telling us to do the same.

"Let me speak to the military, and don't say a word," Wolfgang advised. In his years in northern Mexico, he had developed friendly connections with the powers in government that might come in handy. Once again, we smiled and waved, praying it would work. Some of the soldiers smiled back and waved us onward. Miraculously, we traversed the plaza, but we still had to go through a military checkpoint to leave Ocosingo. We needed to pass as tourists and not geologists because Mexicans believed the CIA routinely disguised themselves as geologists. We found one open *artesanias* shop, and I entered to find a scared elderly woman minding a store of colorful hand-stitched pillowcases. I bought one for an exorbitant forty dollars, just so we could present as tourists. Back in the bus, I held my breath and dug my nails into the pillowcase as we entered the checkpoint. The guards suspiciously surveyed our bus and focused on my pillowcase; satisfied we were tourists they waved us through. My heart hammered all the way back to San Cristobal de las Casas. Along the road I counted more than 115 freshly dug graves.

What would make a sane person foolish enough to enter an area of armed conflict? The Dinosaur Wars, of course. I was still chasing rocks of the K–P boundary for evidence of what really killed the dinosaurs, and in the early 1990s everyone in this hunt was in Mexico.

In 1990 and 1991, a few years before I found myself in Chiapas, two spectacular discoveries had rocked the mass extinction debate. The first was the discovery of a twenty-centimeter-thick layer of two- to four-milliliter-sized glass spherules in a roadcut near the city of Beloc in Haiti.[21] As you might recall, when a meteorite crashes into Earth it can vaporize rocks into a giant dust cloud that will rise into the upper atmosphere. There, the dust cloud condenses into glass spherules which rain back to earth. The discovery of spherules in Haiti triggered a wild rush to find more around the Gulf of Mexico. But this search was not

THE CRATER   75

successful because the Gulf Stream current carried away the sediments throughout this area, as it still does today.

Scientists determined that the best location to search was in northeastern Mexico, which was away from the influence of Gulf Stream erosion. The first successful discoveries of impact spherules in this region came from two outcrops: one in El Mimbral, a small mountain range located one hundred miles west of Monterrey, Mexico, and the other in El Peñón, which is forty miles southwest of El Mimbral. Subsequently, dozens of additional spherule localities were discovered along the same north-south transect over 350 kilometers parallel to northeastern Mexico. We searched for the primary (original) undisturbed glass spherule fallout that might have been deposited first. This was the jackpot of impact localities.

Soon, scientists began a frenzied search in the region to find more impact spherules to determine the age of this meteorite impact. This search led to the second discovery: an impact crater in the subsurface of the northwestern state of Yucatan, Mexico.[22] This crater was instantly interpreted as evidence of the meteorite impact that presumably triggered the fifth mass extinction.

The crater, named Chicxulub (pronounced chick-zoo-loob) after a nearby village, was a large circular structure about 175 kilometers in diameter buried beneath nearly 800 meters of carbonate rocks. The asteroid or meteorite that caused this structure would have been an immense space rock about 10 kilometers in diameter. Imagine this massive rock careening through space, bursting into our atmosphere in a blaze of fire and crashing spectacularly into Yucatan, which was then a shallow sea teaming with life. Today, the impact crater is invisible to the naked eye; if you visit the crater site all you'll find on the surface are acres of flat arid land dotted by scrub brush and palm trees. Scientists detected the subsurface crater based on magnetic anomalies that outline the different rock types of the circular crater.

It's hard to overstate the importance of this twin discovery to the impact theory. Despite the theory's growing popularity and influence in the geosciences, the evidence supporting it was still sparse and shaky. It was also notably lacking the most critical piece of supporting data:

76 THE LAST EXTINCTION

the impact crater itself. So far, the search for scientific support for the impact theory had led impactors to the periodicity hypotheses, which had received a lot of media attention but was largely based on flawed data and statistics, and fantasies like the Nemesis death star. The periodicity hypothesis was supposed to strengthen the impact theory. Instead, its failure took the impact theory down a notch. By the end of the 1980s, the impact theory had lost its novelty and much of its glamour. This was the atmosphere when the twin discoveries of impact glass spherules and the Chicxulub impact crater jolted the impact controversy back to life.

Discovering the massive crater and melt rock glass spherules were the first truly direct supporting evidence of a large impact. The theory was no longer a wild guess, but a testable hypothesis. This major breakthrough attracted scientists from diverse fields from paleontology to geophysics and astrophysics. Unsurprisingly, the popular and scientific press, led by *Science* and *Nature* magazines, proclaimed the controversy was over, the impact theory proven. Many scientists still sitting on the fence were converted into firm believers that the impact was the sole cause of the fifth mass extinction.

Despite my longtime opposition to the impact theory, the new discoveries thrilled me. I remained open to any hypothesis so long as the science supported it, and the Chicxulub crater was an intriguing new piece of geological evidence. This was my opportunity to test the age of the impact spherules and the impact crater. If the crater could be determined to be precisely KPB age, then the impact theory had credibility. I would admit defeat and move on to other topics. However, since my wealth of fossil data pointed to progressive extinctions, I suspected the crater age would not coincide with the mass extinction. Impact proponents viewed Chicxulub as *their* Holy Grail, but it might just as well be *mine* and could yield proof this impact *did not* trigger the mass extinction.

I was champing at the bit to test the impact's age, but Majda was quick to caution me. "Liebling," he started, using our German term of endearment, "there is no need for you to carry on this fight. Dating the crater won't solve anything. They're convinced they've confirmed the impact theory and won't change their minds. You've done enough. Let's enjoy more time together."

THE CRATER   77

He spoke from the delicate position of the abandoned husband. Solving problems was my addiction that easily turned me into a workaholic, although one driven by the joy of discovery. Now, Majda saw a chance to wean me from this mass extinction addiction. He also understood the us-vs-them culture cultivated by the impact theory believers in the 1980s was at a fever pitch, and he didn't want the witch hunters out for my blood.

He had a point. I sighed. "Maybe you're right. For now, I'll sit on the sidelines and see what evidence they come up with."

Majda cocked one eye and regarded me skeptically. He knew who he married.

The popular and scientific press hailed Chicxulub as a surprising breakthrough, but in truth, the discovery of a crater wasn't entirely unexpected. The possible existence of the Yucatan impact crater was proposed in 1981 by geophysicists Glen Penfield and Antonio Camargo, consultants for Mexico's oil company Petróleos Mexicanos (PEMEX).[23] At the time, no one paid much attention to their estimated approximately 180-kilometer-in-diameter circular subsurface structure. After the discovery of the Haiti impact spherules in 1990, Alan Hildebrand, a Canadian graduate student and fanatical impact believer at the University of Arizona, tracked down Penfield and Camargo to discuss their 1981 work on the impact crater, and they published the results.

The existence of glass spherules was not recognized until their discovery in Haiti in 1991. Glass spherules of unknown origin had been observed as early as 1986 in outcrops along the Brazos River in Texas but were ignored. Then I received samples from paleontologist Thor Hansen asking me to identify the forams across the KPB, which I did. Sedimentologist Jody Bourgeois claimed the discovery of an impact tsunami in shallow waters along the Brazos River. This was big news for impactors as the first discovery of impact tsunami deposition. But the forams identified from Brazos River were younger than the "impact tsunami," which was a shallow water storm deposit. My observations angered Jody and Thor, and they withdrew my foram study. Jody's impact tsunami paper became a big winner by 1988. By 1992, Smit turned to Jody's impact tsunami as evidence for Haiti, northeastern Mexico, and

78 THE LAST EXTINCTION

virtually all "impact tsunami," although there frequently was not a shred of evidence.

In 1990, the twin discoveries of the impact spherules and impact crater failed to clarify one critical fact: the age of the Yucatan crater. The mere existence of a crater is not enough to tie an impact with a mass extinction. For there to be the possibility of a causal link, the age of that crater needs to be the same as the mass extinction. Unfortunately, dating an impact crater is virtually impossible. The accepted method at the time was the radiometric dating of glass spherules based on the ratio of argon gas decay $^{40}Ar/^{39}Ar$. But Argon dating doesn't provide a precise enough measurement because argon gas loss over sixty-six million years introduced an age error of 1 to 3 percent, or 0.6 to 1.8 million years. Even today's improved techniques leave an error margin of plus or minus 200,000 years. Relative age dating based on forams is more reliable because the age derived will show for certain whether the crater age is below, at, or above the KPB mass extinction.

This is why, in 1990, Alan Hildebrand asked for my help to determine the crater's age based on forams. I was eager to do so but he only sent me a single sample of limestone from seventy meters above the impact breccia, which dated six million years *after* the KPB mass extinction. I asked him for core samples closer to the KPB and received a baffling reply that the cores had been lost in a PEMEX warehouse fire. How I was meant to complete a comprehensive study was beyond me.

When pressed on these contradictions during a phone conversation, Alan tried to convince me that dating the crater precisely at the KPB extinction was possible based on this argument: *The Chicxulub impact crater is KPB age because this impact caused the iridium anomaly and the mass extinction.* This was classic circular reasoning. There never was any proof that the iridium originated from the impact, or the impact caused the mass extinction. These were the same assumptions parading as evidence going all the way back to 1980.

In his 1991 paper, Alan followed this well-worn path and successively strengthened his claim for the crater's KPB age by repeating this flawed argument, escalating from tentative to conclusive without introducing any

THE CRATER   79

new supporting facts. In the title of the paper, the crater is referred to as "probably KPB age." The abstract of the paper then claimed, "The age of the crater is not precisely known, but a KPB age is indicated." By the conclusion, he confidently asserts "The crater formed at the K–P boundary."[24]

It was a depressing reminder that a powerful aversion to evidence and facts still reigned in my field.

In my office at Princeton, I watched through my window as falling snow piled up along the driveway to the faculty parking lot. It was February 1992, and on the desk in front of me was the new issue of the journal *Geology*, which contained the latest paper by the impact group, and a new sensational claim: that the impact triggered a giant tsunami so powerful it had deposited impact spherules far away in the El Mimbral rock outcropping in northeastern Mexico.[25] The principal author was Jan Smit. I had to hand it to the impactors—their "discoveries" were rarely boring even if they simply co-opted Jody Bourgeois's Brazos River impact tsunami hypothesis, expanded it, and popularized it as their own. But this was a fantastical conclusion by any stretch, no matter who proposed it first. The natural and celestial phenomena they'd dragged into their increasingly convoluted scenario had now run the gamut from flaming meteorites to nuclear winters to death stars to tsunamis. *What was next?* I wondered. *Best not to ask.* This was another electrifying hypothesis based on assumptions without supporting evidence. Predictably, the media loved it.

My phone rang. It was Charles (Chuck) Officer, the last survivor of the anti-impact cabal beside me. We briefly talked about Smit's paper, and then he got to the point of his call.

"We've got to check this out," he said. There was nothing Chuck liked more than getting the real story behind impactor's wild theories.

I was less enthused and told Chuck I wasn't sure I wanted to continue research on the KPB mass extinction. But Chuck was insistent. "You've got to continue," he urged me.

When he wanted to, Chuck could lay it on thick. He told me I was the only one who could take on the impactors. That if I didn't help to expose their faulty science it could set the field back decades.

80 THE LAST EXTINCTION

I let out a nervous laugh, trying to lighten the mood. "Chuck, you're overdoing it. I'm just one person. How far can I go against virtually the entire US science establishment? I'd be committing career suicide."

But Chuck wouldn't relent and ended with a final appeal, telling me, "You owe it to science."

I knew he was buttering me up, and I had to admit it was rather effective. Finally, I relented.

Chuck's plan was for me to organize an international field trip to northeastern Mexico, to study El Mimbral. There, we would examine the section and see how likely the Smit scenario was.

I could see why Chuck had been a successful businessman: He was dogged and persuasive. Listening to him outline the expedition with half an ear, I remember looking out the window and seeing the lights of nearby university buildings start to come on in the rapidly descending darkness. I told Chuck we'd continue the conversation some other time, bid him a cheery "Ciao!" and hung up before he could protest further.

On my way home, I mulled over Chuck's field trip proposal. I wanted to see this locality for myself, but I didn't know anybody who worked in northeastern Mexico, spoke next to no Spanish, and had no clue how to arrange a field trip for a bunch of international professionals. And why should I be the one to arrange the trip?

Over dinner, I told Majda about Chuck's call. I chose my words carefully, knowing Majda wouldn't support my return to the Dinosaur Wars. I presented the trip to him as an opportunity to see the section myself and then decide whether to continue. Majda smelled a rat.

"If you go, it's clear that you'll continue your work, which will be endlessly stressful for both of us."

I consoled him that he was right. I would stop and take his reservations seriously, while knowing all too well I was preparing to reenter the fray.

After a restless sleep, I woke the next morning jittery with anticipation. That familiar thrill of launching a new project, a new foray into the unknown, danced in my chest. Today, I would step off the sideline and begin to organize the field trip to Mexico. I hurried through breakfast and made my way to my office in record time. First, I called

Xavier Hellenes, a fellow Stanford graduate student who had taught at the University of Nuevo Leon in Linares, which is in the vicinity of El Mimbral. From there, the pieces fell into place with astonishing ease. Xavier connected me with the university's department chair, who proposed two department members to help organize this trip: Professor Wolfgang Stinnesbeck, a German paleontologist, and Guadalupe (Lupe) Lopez-Oliva, a teaching assistant. Both quickly came onboard and organized the field trip. I couldn't believe my luck.

Everything was set up, with one important exception: Neither Wolfgang nor Lupe could find the El Mimbral outcrop. Smit, Alvarez, and others had given little information of Mimbral's location, and the coordinates were imprecise. The landscape surrounding the Mimbral Valley quickly turns wild, with few roads or notable landmarks. Without additional guidance, our odds of finding Mimbral were slim. No outcrop, no field trip.

I knew who could help . . . if he wanted to. I picked up the phone and called Walter Alvarez. When I congratulated Walter on the splendid discovery of the Chicxulub impact spherules at El Mimbral, he was pleased.

"I'd love to see the section," I told him and asked for specific location information, like landmarks that would help us find the outcrop.

Walter's mood soured. "The evidence for the impact-tsunami is strong, Gerta," Walter said. "Just let me rest on my glories for a while."

He refused to give me any location information, and I could understand his reluctance. My research had the pesky habit of spoiling their best stories. The tsunami interpretation was as bogus as the impact theory itself, and he knew if I found any incompatible evidence, I wouldn't hesitate to make it known. We exchanged frosty goodbyes.

In late March 1992, fourteen scientists from France, Germany, and the US arrived in Linares for the field trip. Linares at that time was a small nondescript city with a single paved main road and a small plaza with two hotels. We were a lively crew, with more than a few characters among us, including Chuck Officer. I was most drawn to the company of Wolfgang, who organized our ground game in Mexico, and his Swiss friend Thierry Adatte, a sedimentologist from the University of Neuchâtel, Switzerland. Although opposites in mannerisms, appearance, and approach to science,

82 THE LAST EXTINCTION

Wolfgang and Thierry were close friends and kept up an entertaining odd couple banter. At six feet three inches, Wolfgang was an easygoing giant with a shock of blond hair thinning on his forehead, his face always lit with a carefree grin. Thierry was shorter with a high forehead and receding dark hair, more subtle and subdued than his gregarious friend. I liked them both immediately.

Finding the El Mimbral outcrop was difficult. After Walter Alvarez had rebuffed my request for location information, Wolfgang received some information from a PEMEX geologist, which put us in the right valley but without coordinates. We split up into groups and searched systematically all day long, cutting through the thick underbrush of scrappy woods, trying to avoid snakes, scorpions, and wild pigs. We were frequently stopped by stern Mexicans on horseback with long rifles glinting in the sun. They were guarding a nearby marijuana cultivation area and tasked with keeping people from discovering it. As we continued our search, they trailed us, ominous figures in the distance. By late afternoon we ran into another Mexican with a rifle on horseback. Wearily, we pulled out the creased and dirtied photo of El Mimbral. Somehow, miraculously, this time was different. Seeing the photo, the man nodded yes, it was just a couple of hundred meters ahead of us. Overjoyed, we found this large, beautiful exposure and ran up and down the rock face like headless chickens.

It was ideal for study. An exotic three-meter-thick rock exposure that rose up like a wedding cake stacked with three distinct units, each one meter thick, that infilled a submarine channel scoured by currents and was about 150 meters wide. All deposition occurred in Cretaceous age sediments below the KPB. In ascending order, the sediments of this channel contained impact spherules, sandstone, and thinly laminated marine sediments, among other characteristics. It was clear to see that the KPB mass extinction and iridium anomaly were above this "wedding cake." Our task was to find out what really happened in this rock sequence and how it fit into the larger picture of what we knew to be happening in the years leading up to the mass extinction.

It was a relief to finally have some impact theory evidence to examine. Proponents of the theory clearly felt vindicated by the discovery of the Chicxulub crater. There it is, they pointed out, proof of a large impact;

THE CRATER 83

the smoking gun that finally proved their hypothesis! But putting forth the crater as evidence also meant that it could be tested, and that opened it up to refutation. The logic behind their argument was simple: An asteroid struck causing a sudden and severe extinction scenario. For this to be true, therefore, the following conditions had to be met: The impact crater, impact spherules, iridium anomaly, K–P boundary, and the mass extinction must all have occurred at precisely the same time. If the evidence didn't line up, it falsified the theory.

And that was already a problem for the impactors, because all the impact spherules in northeastern Mexico were found in Cretaceous sediments well below the KPB, which is incompatible with the presumed KPB age of the impact crater and the impact theory. That was why, to reconcile the older spherule age with the presumed KPB impact, Smit proposed a giant impact-generated tsunami that mixed the spherules into Cretaceous sediments. In this tsunami scenario, the impact crashed into Yucatan and spherules rained from the impact's dust cloud within hours. Then a tsunami created by the impact transported the spherules into the deep and an earthquake caused sandstone to slide downslope on top of the spherules. The ending phase of the tsunami deposited the fine sand and mudstone. At last, the iridium settled from the sky upon the mass extinction. It was an imaginative scenario but incompatible with the evidence on the ground.

Our investigation of the evidence was much more plausible. Characteristic marine deposition occurred over a long time period with no evidence of tsunami disturbance.

At the base of El Mimbral, where we first tested Smit's tsunami hypothesis, we found the one-meter-thick interval to consist of two impact spherule-rich layers with shallow water debris that had been transported downslope in a submarine channel by currents and redeposited in deeper water. This is a common occurrence of near-shore erosion and transport. Between these two spherule layers is a white twenty-centimeter-thick limestone. This was the first warning that the impact-tsunami scenario was false. Limestone precipitates in quiet waters at a rate of approximately 2.5 centimeters per 1,000 years, which means the twenty centimeters would have accumulated over about 8,000 years.

84　THE LAST EXTINCTION

If there had been a tsunami, we would have seen mixed and jumbled sediments, just as Jody Bourgeois had found at the Brazos River. But there was such evidence of a tsunami here. That means the spherules at El Mimbral were not transported from far away and mixed into other sediment layers. This was normal quiet sediment deposition, and the spherules created from the Chicxulub impact fell into these layers well before the KPB. The impact spherules predated the mass extinction.

This wasn't the only evidence we found in Mexico that contradicted the impactor's hypothesis. We also observed that the composition of the middle one-meter-thick sandstone was too coherent to have been involved in a tsunami deposition. Had it been subjected to such forces its structure would be inherently chaotic.

Yet more damning evidence against the tsunami interpretation was evident in the top one-meter-thick interval of millimeter-thin laminated fine sand and mudstone. These thin layers are intensively colonized by seafloor burrowers and grazers that feed on organic-rich sediments. Laminated sediments are deposited in quiet water over a long time. We estimated that sedimentation in this interval occurred at about three and a half centimeters per 1,000 years or a total of approximately 28,000 years to deposit one meter. You can't posit that the tsunami was powerful enough to shuffle the location of the spherules and believe it didn't disturb the life living on the seafloor, of which there was abundant evidence.

In any case, laminated sediments are incompatible with high-energy wave action, especially tsunami waves. Likewise, a colonized seafloor can't exist during a tsunami. It was like saying a tornado could destroy a town while at the same time leaving all the wildlife untouched.

The evidence we saw at El Mimbral was so unremarkably clear it raised uncomfortable questions about the arguments used to support the impact theory. Was the impact-tsunami scenario invented to save the impact theory? Was contrary evidence omitted because it was incompatible with tsunami deposition?

I believe so. When I asked Smit why he ignored the fossils in the laminated sediments and the limestone between the spherule layers, he said there were no fossils, and the limestone was injected into the impact spherule layer by an impact-triggered earthquake. This assertion

THE CRATER   85

was difficult to take seriously, describing as it did geological movements never seen before. How could an earthquake inject a thin limestone layer into a spherule-rich deposit over 150 kilometers?

In retrospect, this moment seems to have been a turning point in the science of impact theory. It was one thing to put forth fanciful theories like the Nemesis star, or to make speculative leaps like the insistence that iridium found at the KPB *must* have had an extraterrestrial source. But now we were dealing with hard evidence. It was worrisome that one of their key pieces of evidence required true believers to engage in logical contortions with no precedent. It suggested that fealty to their theory was guiding their conclusions, not a dispassionate appraisal of scientific facts. For the impactors, conjecture seemed to have hardened into belief, and thus every piece of evidence had to be interpreted to support the belief. But even in light of the subjective quality of some geological

(A) At El Peñón in northeastern Mexico, the Mendez Marl Formation is above a twenty-five-centimeter-thick impact spherule layer, followed by fifteen centimeters of sandy limestone, and followed by a second impact spherule layer 10cm above. Limestone precipitates slowly over thousands of years, which rules out the idea of a giant impact crashing into Yucatan and generating impact-tsunami waves. The evidence supports normal seafloor life scavenged by invertebrates. (B) Most curiously, we found scavengers within J-shaped burrows infilled by impact spherules. (C) Blow up of same burrow with impact spherules.

86   THE LAST EXTINCTION

evidence, there was no reason to believe that what we were looking at showed evidence of a tsunami. We had all seen outcrops showing normal sediment behavior, and we had all seen the geological results of tsunamis. El Mimbral looked nothing like the latter. Whatever was guiding the conclusions of the impactors, it wasn't science.

My suspicions only grew later that year. In late May we returned to Mexico to visit El Peñón, the other location in which impact glass spherules linked to the Chicxulub fallout had been discovered. There we found exactly the same story we'd uncovered at El Mimbral. In both localities, a one-meter-thick glauconite and spherule layer unit containing a twenty- to twenty-five-centimeter-thick sandy limestone was discovered at the base of the sandstone complex. El Peñón was on a more gradual slope, so there was less deposition between the layers. But there was still no evidence of a tsunami, and the spherule layers were clearly located below the KPB.

By this time, our critics had responded to our analysis. Smit doubled down on the tsunami theory and suggested the limestone layer separating the spherules from the KPB was due to a large-scale tectonic disturbance. There was no evidence of this. They also claimed the burrows we had found that could not have survived a tsunami were too few to be of significance, were "fluid escape structures." Other novel explanations were put forth to explain how a fairly conventional assortment of sediment was actually hiding a cataclysm. But none of these ad hoc arguments has been supported by evidence, nor can these explanations account for the evidence based on field and laboratory observations.

For Wolfgang, Thierry, and me, it became increasingly obvious what really happened sixty-six million years ago had nothing to do with an impact tsunami or the impact theory. Our exhilarating discoveries reconnected me with the spirit of my early adventurous life, which now added competitiveness and venturing wherever I had to go to prove what really happened during the fifth mass extinction. And best of all, our growing camaraderie added fun and joy, as we naturally settled into collaboration that lasted for decades.

7

UNDER *the* REVIVAL TENT

SWISS MOUNTAIN PEOPLE ARE SAID TO HAVE "HARD HEADS." My stubborn optimism, perhaps bordering on naivete at times, has been both a superpower and an Achilles' heel throughout my career. It has given me the courage and conviction to attain great success but has also blinded me to potential disaster.

One way my hard head has occasionally led me astray is by convincing me to place too much trust in some of the scientific systems and organizations that governed my field. Over the centuries, scientists have developed excellent methods and institutions by which to evaluate claims. But scientists are human beings, and their motives can be murky, and not always aligned with the rational pursuit of truth. I was to discover the dangers of misplaced trust in October 1993, when I made one of the worst blunders of my career.

It all began when I was asked by Buck Sharpton, the NASA scientist with whom I'd briefly communicated about access to PEMEX core samples, to participate in a workshop in Puerto Vallarta, Mexico, to discuss a proposed plan to drill into the Chicxulub impact crater for sediment samples.

I was surprised to get his call and also surprised by the lateness of the invitation. The workshop was to be held sometime in the next few months: "Why me?" I said. "And why such a late workshop invitation?"

88   THE LAST EXTINCTION

"Because we need to diversify," Buck replied unabashedly. "NSF funded workshops must include dissenters like you."

"What an honor to be chosen as the token other at your workshop," I replied.

Buck laughed. "Just think of it as a free trip, all expenses paid, at a great hotel by the beach. You'll enjoy it."

Buck seemed determined to sell me this cost-free invitation. Had I really become the token outsider with a seat at the table just so NSF would approve funding for the Chicxulub drilling workshop? Surely, the impact heavies wouldn't allow me to have a say. On the other hand, having followed the publicity of this impact crater-drilling plan, I was all in. But the actual Chicxulub impact drilling was still lightyears away and not expected before 2001 or 2002. Accepting this invitation could improve my chances to eventually obtain samples. That hope alone clinched it.

But Buck wasn't finished. Before I hung up, he told me we had more business to discuss. He wanted to know if I could organize and lead a field trip for scientists and journalists through northeastern Mexico before the Lunar and Planetary Institute's upcoming conference in February 1994.

I almost dropped the phone. He was talking about LPI's Snowbird III conference. After my disastrous appearance at Snowbird II, I would have thought I'd be the last person they'd want.

"Why me?" I asked. "Why not one of your own true believers in the impact theory, like Jan Smit?"

"We want a different perspective, as you and your team expressed in the *Geology* paper,"[26] Buck answered. He was referring to the paper that described our evidence from El Mimbral, which showed the Chicxulub impact predated the KPB and that there was no evidence of an impact-generated tsunami event as the impactors claimed.

I was skeptical. Buck was describing something very close to the open-minded, fact-based give-and-take that had been so absent from my previous encounters with impact theorists. After so many years of impact theorists attacking and denying my discoveries, I asked Buck why I should believe that this conference would be any different.

Buck assured me he was sincere, and that this was a chance for me to convince those with open minds.

UNDER THE REVIVAL TENT   89

My mind reeled. Was Buck serious about wanting to hear both sides? Or was he just setting me up in front of the firing squad? It was fair to say that scientists at LPI, a unit of NASA, harbored no doubts that a large asteroid impact caused the KPB mass extinction. The field trip would be no place for skeptics, and even less for a skeptic leader.

My common sense told me stay away. My "hard head" told me to take them up on the offer.

"I hope you accept," Buck said. "You're our one choice to lead the field trip."

I very much doubted it.

After much hand-wringing and consultation with Thierry and Wolfgang, who opposed the idea, I accepted Buck's invitation to lead the field trip. It was a terrible decision. Even now, when I am well into my seventh decade, I get angry when I think of it. Why hadn't I learned my lesson? But then I realize that to decline would have been against my nature. I could never have turned down the field trip invitation, just like I couldn't help myself leaving that highway in Laos. I wasn't going to walk away and let a bunch of middle-aged male scientists deride me as a scared rabbit. I'd never been a scared rabbit in science or in life.

Once I had made up my mind to lead the field trip, it wasn't long before my stubborn optimism turned into wishful thinking. This, I convinced myself, was a once-in-a-lifetime opportunity to demonstrate the evidence in the field! I would walk up to the sediment layers in the light of day and point at the irrefutable proof. How could anyone deny the evidence if it was right in front of them? My mind started conjuring vainglorious fantasies. I imagined the field trip group gathered around the El Mimbral outcrop, and someone crying out, *Wait a minute! The impact-tsunami scenario is not supported by the facts on the ground. The Chicxulub impact may not be KPB age after all!* I was a sucker for believing that real evidence could sway the debate. A girl can dream.

On my first night in Puerto Vallarta, I descended the wide curved staircase into the hotel lobby to join the Chicxulub workshop welcoming reception. It was six in the evening, and outside the large windows, the sun cast a warm light on the azure waters of the Pacific Ocean. Feeling awkward, I paused halfway down the staircase to survey the crowded

90 THE LAST EXTINCTION

lobby. Close to the bottom of the stairs, I spotted a tight group of about three dozen men with suntanned ruddy faces, dressed in baggy jeans, bland T-shirts—and some with safari vests—beer bottles in hand, talking and drinking. Unmistakably geologists. You can identify them anywhere.

One man who didn't quite fit the geologist mold separated from the group, two glasses of wine in his hands, and approached the staircase. He was tall and slightly built, wore a dark gray suit, and sported a shock of gray hair. A curious and amused look played on his face as he navigated his way toward the stairs, occasionally looking up and smiling. I turned around to see at whom he was smiling; there was no one there. Was it me? I'd never met the man.

At the bottom of the stairs, he gestured toward me, lifting both glasses of wine, and offered: "You look like you could use one of these."

Grinning, I walked down the steps and introduced myself.

"I recognized you," he said, smiling, as he handed me a glass of wine, which I accepted gratefully. "I'm Bill Ward, University of New Orleans. Tell me, what were you thinking on these steps overlooking all those geologists?"

"What the heck am I doing here?" I replied.

He burst out laughing, "That's exactly what I was thinking too."

Sensing an ally, I brought up a mystery that had troubled me for years—what had happened to the missing PEMEX Yucatan core samples. I told him that I thought the story that the cores had burned in a warehouse fire was baloney and made up so nobody else could investigate the cores.

Bill looked at me surprised. "I know for a fact this isn't true," he replied. "I have the half-split cores in my lab at the University of New Orleans." Bill explained that he inherited the Mexico cores from his predecessor, Arthur Meyerhoff,[27] who consulted for PEMEX during the 1970s. In fact, he had recently been told by LPI and their Mexican counterparts that they wanted the cores returned immediately after the workshop on Chicxulub drilling.

So, I was right. There never was a warehouse fire that destroyed the KPB cores, as Alan Hildebrand had told me in 1991, when he asked me to do a foram analysis of other PEMEX samples for him. Had Alan

UNDER THE REVIVAL TENT 91

been misinformed, or was he withholding those samples for fear they would show that the impact preceded the KPB? I had reason to believe it was the latter, as I had also heard that a previous study by a PEMEX foram consultant had revealed Cretaceous sediments above the impact crater breccia.

"Wow!" I exhaled. "Too bad. You just resurrected the cores only to have them hidden away for good where nobody can study them—at least not skeptics."

"That's not necessarily true. They are still in my lab—what do you have in mind?"

My jaw dropped. "Really? Could I interest you in studying the cores with me and my collaborators before turning them over to Mexico?"

"I would be honored," he replied with a fiendish smile, and we clinked our wine glasses.

Suddenly the world looked rosy as I surveyed the crowd of impactors.

Almost forgotten in the excitement of studying the rediscovered PEMEX cores was the upcoming announcement of who would lead the Snowbird III field trip with its seventy-plus international participants. As it turned out, I still needed to audition for the role. The other candidate was José Longoria, a native of Monterrey, Mexico, who was teaching at Florida International. He had been with the Alvarez group on field trips in northeastern Mexico and seemed well-positioned to lead this field trip.

Longoria and I met over lunch in the hotel restaurant with Buck Sharpton, Graham Ryder, and a couple of others from LPI who questioned Longoria and me as to how we would organize this field trip. Longoria proposed to shuttle the participants by bus from Houston to Monterrey, Mexico, and over two days visit a dozen outcrops. I proposed to fly the group to Monterrey and visit only the three most critical outcrops, to leave plenty of time for viewing, discussions, and taking samples. When LPI scientists voted for my program, Longoria turned red-faced and called me a "Danish bitch."

There were audible gasps around the table. Then a shocked silence. I watched the faces of my colleagues. Their eyes were downcast, they seemed to be barely breathing—all waiting for my reaction.

92   THE LAST EXTINCTION

I burst out laughing, declaring: "Wow, as a Swiss I've never been called that before."

The spell was broken, and a sigh of collective relief expelled the tension around the table. In the awkward silence that followed, Longoria stood up and left.

I was now the field trip leader, saddled with the dubious honor of planning a massive international expedition for six dozen of the world's most hardcore impact believers.

My field trip team included Wolfgang Stinnesbeck, Thierry Adatte, and me. We had organized the Lunar and Planetary Sciences (NASA) field trip for over seventy scientists and more than thirty people from press and film crews, and arranged transportation, housing, and meals. Buck Sharpton, strangely, was not in attendance, but outside of this surprise, everything was in order. The governor of the State of Nuevo Leon invited our motley group of geologists to a sumptuous dinner followed by folkloric dances. It had the potential to be a great field trip, but I also felt a sense of nervous dread.

We shuttled 110 people in minivans to the Mimbral outcrop where media people set up tents for filming and interviews. The outcrop stretched about three hundred meters along a steep hillside with rock numbers painted red by the impact team of 1992. Most scientists mingled around the fringes of the media tents hoping for interviews. Barely a dozen scientists briefly climbed up the outcrops for a brief look. Only three to four serious scientists from the US studied the outcrops and collected samples.

At 10 a.m., Bob Ginsburg from the University of Miami took the bullhorn and announced LPI had appointed him as the field trip leader. "I'm a marine carbonate scientist who loves the impact theory," he said, "but I don't know anything about the geology of rocks and fossils." After Ginsberg's charming admission of his ignorance on the relevant subject matter of the trip, Smit claimed the bullhorn with a big grin, glancing my way with great satisfaction—like a kid who just stole the candy.

My mind reeled. For planning and organizing this field trip, Buck Sharpton promised open discussions from field evidence. We had just discovered the impact spherules in northeastern Mexico predated the

KPB mass extinction. Smit claimed the usual impact-tsunami mixed with sediments of KPB age. Of course, there wasn't any tsunami evidence; it had been made up by Smit and his collaborators from the very beginning to support the pre-KPB age of the Chicxulub impact. And now Smit and Ginsberg had taken over *our* field trip with not a word acknowledging our work or even existence, let alone our weeks of hard labor in preparations at our personal costs.

How could I have been conned into believing there would be an honest assessment of the outcrop in the field? How could I look at my new collaborators, who had put their trust in me, in the eyes? They had organized and run most of the field trip with gusto, while at the same time being skeptical that we would be cheated.

My thoughts turned to Buck Sharpton, who organized the field trip from NASA's Lunar and Planetary Sciences. This betrayal must have something to do with why, before the onset of the field trip, Buck's role had suddenly ended without a word, and why he didn't attend the field trip. He'd struck me as a sober impactor with integrity. He must have known that Jan Smit, a loner with an aggressive personality who didn't get along well with many of his colleagues, was always a poor choice for a field trip leader. Smit was not equipped to organize and lead a successful field trip. My team, however, had the capacity, knowledge, and skill to lead a large international field trip. But apparently, Buck's integrity was too much for Smit, who eliminated him. The field now belonged to Smit and Ginsberg, leaving no question that the impact tsunami was just "truthiness."

My team never had a chance to present the real science at El Mimbral and expose the errors and misrepresentations that propped up the tsunami "theory." We were prepared to present these facts to the scientists and journalists in attendance, to point at the evidence and show them what it indicated. But Smit and Ginsberg pushed us aside to present only their interpretation.

We had wasted weeks preparing for the trip. We'd organized it in good faith. But it turned out to be a monumental deception. I had the nightmarish sense we'd been set up as the sacrificial lambs at the altar of Chicxulub. It was the field trip from hell.

94 THE LAST EXTINCTION

Field trip leaders, from left to right, Gerta Keller, Wolfgang Stinnesbeck, and Thierry Adatte at the LPI sponsored Snowbird conference (Feb. 4–8, 1994) in northeastern Mexico to examine evidence from the Chicxulub impact. It was the field trip from hell.

I have no doubt that one of the main reasons we were sidelined was because of the presence of media on the trip. This circus dismayed me almost as much as Ginsberg and Smit's hostile takeover. The press contingent was huge, including journalists, cameramen, film crews, biographers, and science history writers all traveling in their own buses. When we reached Mimbral, they set up tents in which they could film interviews with chosen scientists. Other scientists hung around the fringes hoping to grab a few minutes of fame. Many of the field trip scientists scrambled for interviews so that they could share the exciting impact-tsunami story with journalists. Ironically, these interviews were often conducted right on top of the outcrops that contradicted their theories. But who could tell the difference? The journalists had never studied a sediment layer or seen what evidence of a real tsunami looked like. They weren't

UNDER THE REVIVAL TENT 95

trained in sediment analysis, and neither were many of the scientists. Two of the invited experts, however, did have the expertise to notice something wasn't right with the tsunami story: sedimentologist Don Lowe from Stanford University and paleontologist and trace fossil expert Tony Ekdale from the University of Utah.

While the tsunami advocates were talking to the media, I noticed that Lowe was carefully examining the outcrop. He concluded that the tsunami interpretation may be premature, but otherwise kept his impressions close to the vest. Ekdale was less discreet. As a well-known expert in marine fossils, burrows, tracks, and feeding traces, he noticed many of them at the El Mimbral outcrop. With excitement, he shouted his discovery down to the field trip participants milling around the media tents.

"Hey, wait, there is evidence of marine life. How could animals have lived on the seafloor during a tsunami?"

The whole group fell silent. I held my breath. Amidst my gloom, I felt a glimmer of hope. This was the moment I'd imagined months ago, when science and evidence would overcome invention. Could this be the turning point? Could Tony possibly inspire them to listen and reason?

In Tony's momentary silence, Smit shouted back: "You don't know what you are doing; those are not fossils but just scratches from your hammer."

There was scattered laughter. Then everyone turned back to the media tents. No one took a closer look at what Tony had found. Nobody was interested in the fossil evidence Tony had discovered and its game changing implications. Humiliated, Tony suffered the rest of the trip in weary silence.

After the field trip, the Lunar and Planetary Institute unit of NASA held its conference on *New Developments Regarding the K–P Event and Other Catastrophes in Earth History* in a large conference hall within LPI's research building in Houston. It didn't matter that we no longer convened in the high altitude of Utah's Rocky Mountains—everyone still called it Snowbird III. The LPI building was a modern concrete structure of clean lines and right angles, but the activity within was better suited to a big-tent religious revival.

96   THE LAST EXTINCTION

In the packed auditorium of about 300 scientists, I watched as impactors ascended to the podium on stage to exclaim into a microphone: "I believe in the impact theory! I believe in the impact tsunami!" With each "I believe!" confession, the audience of impactors clapped wildly, shouted approval, and banged on the small desktop attached to their seats. All that was missing were ecstatic cries of "Amen!" and "Testify!" to complete the scene. This was true believer country.

The experience was surreal—hilarious in its absurdity but also scary in its fervor. I sat in the front row with a small group of fellow skeptics, trading looks of astonishment and amusement. None of us could believe what we were seeing. How could smart, supposedly logical scientists behave like religious fanatics? I could never understand the reason for this spontaneous public confession of true belief at a science conference, especially because 95 percent of the scientists in attendance were already true believers. This is still the most unbelievable weird experience I've ever witnessed in my science career.

This couldn't happen at any other geological conference. Snowbird III was, for impact loyalists, an essential place to be and be seen. But it was a relatively small conference with only about three hundred attendees. In contrast, the large geo conferences held by the GSA, AGU, and EGU each attracted more than 10,000 participants. Few of the "silent majority" bothered to attend the NASA-funded "Snowbird" conferences, which were designed to celebrate, promote, and publicize the impact theory. As a result, these meetings fostered a club-like atmosphere of groupthink that punished even the slightest deviation from dogma. By the 1990s, NASA was playing an outsized role in supporting the impact theory, an outgrowth of its expanding mission to protect Earth from the threat of extraterrestrial objects.[28] It had been a mutually beneficial marriage from the beginning and a win-win strategy for both sides.

I usually enjoyed small conferences and looked forward to reconnecting with old friends, but I dreaded Snowbird and NASA LPI meetings. These events dredged up awful memories of my public humiliation at Snowbird II and made me wonder what unpleasant surprises lay in store. But I had to be there. The results of the blind test of my El Kef study, which I had presented at Snowbird II in 1988, were finally going to be released after an inexplicably long delay. There was no way I was going to let the presentation of the analysis happen without me. I needed to be there to explain the results to my colleagues. Backing out was not an option.

Despite my disappointment over the field trip bait-and-switch, I was in a good mood going into the conference because Bob Ginsburg, who had been appointed the "impartial" blind-test mediator, had already shared the conclusions with Smit and me. They were strongly in my favor. They supported my contention that the site showed a progressive mass extinction and survival pattern where large specialized species went extinct first, while small simple species survived the mass extinction for some time. None of the four blind test results supported Smit's theorized extinction pattern which posited that all species were instantaneously made extinct by the impact, save one single surviving species.

Ginsberg wanted Smit and me to present the pros and cons of our arguments at the conference. Whenever I asked him about the schedule for the release of the tests, he grew evasive, as if he were buying time. He knew that the blind test results pointed to an extinction pattern that was incompatible with a sudden, impact-related mass extinction. I suspected he was protecting Smit and the impact theory by delaying the release of the report. This Good Old Boys Club and their games exhausted me.

Apparently, I exhausted them as well. As I wandered alone among throngs of scientists during coffee breaks and poster sessions, scientists frequently turned their backs to me when I approached or closed their ranks to whisper. I stood awkwardly outside their circle, clutching a paper cup of lukewarm tea. Conferences are cliquey affairs to begin with, but this was next level. The few impact skeptics I recognized didn't dare being seen talking to me or even standing next to me for fear they, too, would be drawn into the cauldron of the witch hunt that engulfed me. At other conferences, members of the silent majority—the skeptics—often congratulated me for my perseverance and courage to speak up, but they rarely showed up at impact meetings. At the small impact meetings, the participants were overwhelmingly male and jockeyed for speaking rights to regale their group with updates on their projects favoring the impact theory.

On the second day of the conference, it was time for my friend and Princeton colleague Norman MacLeod to give his talk. I sat beside Norman in the front row, and before he rose to speak, he turned to me and said, "Watch me, I'm gonna have a little fun."

Norman is a funny character, a devilish joker, and sometimes a loose cannon. He could be moody, but also hilarious and fun. I had no idea what he was planning. *Oh dear*, I thought as I watched his slim figure, clad in his signature tight blue jeans, vest, and cowboy boots approach the stairs.

He ascended the podium like a preacher to the pulpit. Then he stood there for an uncomfortably long time, surveying his flock. A small, slim, curious figure with curly shoulder-length graying hair, he stood completely still. The room fell silent. Then in the cadence of a Southern preacher, Norman let loose a thunderous, "I BELIIIEVE!"

The audience startled. What was this skeptic doing? Had he converted? I held my breath.

Norman roared again: "I BELIIIEVE in the impact theory!" I turned my head to assess the audience reaction. Confused silence. Then applause kicked in, at first hesitant clapping by a few, then more.

"I BELIIIEVE in the impact-tsunami!" Norman bellowed. Now the crowd let loose with rapturous applause as Norman whipped them into an ecstatic frenzy with repeated "I BELIIIEVE!"s.

Norman grew up in Texas, and his mother was a revivalist. He must have witnessed many pastors in action because his performance was riveting. Our small group of skeptics laughed so hard that tears ran down our faces. At the same time, it was frightening and sad to see such a large audience of scientists abandon all reason in the thrall of belief. No one else understood the joke. (Dick Kerr, a writer for *Science* who was sitting several seats over in the front row, later reported Norman's performance as a serious change of heart.) Norman followed his sermon with his planned talk, in which he cautioned that the impact theory alone failed to account for the mass extinction.

UNDER THE REVIVAL TENT   99

I, of course, did not please the crowd with my talk. I used my time to again point to the evidence that should force them to reconsider the impact-tsunami scenario. I was greeted with hisses and boos and a continuous din of dissent that drowned me out. It was another childish act of bullying disguised as academic dissent. Still, I held out hope that the results of the blind test, once released, might yet make the conference a success.

The conference was scheduled to conclude on Saturday at noon. But on Friday, a massive snowstorm began on the East Coast that was predicted to last for days and lead to airport closures. Ginsburg encouraged all the East Coast travelers to leave on Friday afternoon so they wouldn't be stranded in Houston. This made me uneasy. What about the blind test results? Who would be there to guide the other scientists and journalists through their implications? I didn't want to leave before results were released.

By late Friday I fretted over whether to leave or stay. Ginsberg assured me repeatedly there would be no release of the blind test results on Saturday. Smit told me the same story. I relented and departed on the red-eye back to the East Coast as the snowstorm threatened any later escape. It turned out Smit deceived me from the get-go and scheduled his release of the blind test results for Saturday morning. While I was sleeping back in Princeton, Smit—with great fanfare—presented the results and announced to the remaining scientists and journalists that the blind tests had confirmed his theory and proved me wrong. There was no one there to dispute him.

When I found out about this deception later that day, I was so angry my hands began to shake, and I felt a jittery tingling sensation that swept my body from head to toe. It was bad enough that I wasn't there to explain the data, but I also heard Smit had told attendees that I'd left the conference early because I was a sore loser who couldn't face defeat. I had been wrong on the blind test, he said, wrong on the Mexico field trip, and couldn't face reality that my life's work was in vain.

Smit's accusations were false, absurd, and even laughable, but they still hurt and made me wish I could crawl into a hole or move to another

100 THE LAST EXTINCTION

planet just to leave this insane world of impactors behind. The truth is that after two weeks immersed in a constantly hostile environment during the field trip and then at LPI (Snowbird III), my nerves had frayed. I began to wonder if any of this was worth it.

It took me a few weeks to piece together how and why the results of the blind test from the 1988 Snowbird II conference had been withheld by Smit and Ginsberg at Snowbird III. They had always been strong impact supporters, and contrary results of the blind test had been withheld since before 1988. Ginsberg had told me in private, the four blind tests confirmed the data—including mine—but Smit's interpretation didn't pass the test. This was the reason why Ginsberg refused to publicly release the blind test results at Snowbird III, calling them "difficult to interpret" and "best to discard the blind test." In short, it was just another way of saying "Ginsberg and Smit were wrong." Ginsburg admitted this much to me on Thursday, the day before the snowstorm.

It was yet another violation of good science in favor of a predetermined result. The blind tests had been performed with impeccable objectivity: The four blind testers worked independently without knowing each other. None of them could discuss their data. They were given specific instructions to count the total number of species above and below the KPB. They noted the time and size each species disappeared before and after the mass extinction. The census results were clear, with no manipulative counts or calculations. The only problem from Smit and Ginsberg's perspective was their theory didn't support the data. Still, they knew the results couldn't be delayed forever, and so they adopted a different strategy.

Ginsberg departed that Friday afternoon to escape the snowstorm but reassured me there would be *no* blind test. I didn't believe his word. By evening I asked Smit once again when the blind test would be held on Saturday morning, the last day of Snowbird III. He assured me there would be no blind test. I didn't believe his word either. But now I had to choose between yet another lie from Smit and wait in Houston to find out the truth, while getting stuck in the East Coast snowstorm for days. I chose to depart on the red-eye special to New York.

Smit deceived us all. On Saturday morning he presented the blind test results, which he had fudged and manipulated at the LPI conference in the absence of Keller. He merged the four separate blind test results to total fifty-nine species. He fudged twelve species as uncertain, which left forty-seven species. He fudged another five species he didn't like, which left a doctored dataset of merely forty-two species, with the rest of the species suddenly extinct, except the hardy survivor *Guembelitria cretacea*. In other words, only one species survived the KPB mass extinction.

So, Smit fudged the blind test results at Snowbird III on the last day of the LPI conference in Houston when I was absent due to the snowstorm that closed airports and stranded travelers for days. Under these circumstances I took the chance that Smit may not have lied this time. I was wrong.

Without me there to refute his absurd erroneous results, the deception spread unchecked. Employing the same unsavory, media-savvy tactics that Luis Alvarez had perfected, the impact camp manipulated the press to cement their victory. Just after the conference, Smit wrote an opinion article for *Nature2* in which he proclaimed himself the winner of the blind test.

Ever the dutiful servant, Dick Kerr, the same reporter who couldn't spot Norman MacLeod's parodic preaching performance, announced Smit's triumph over the blind test reveal as the most important thing to happen at Snowbird III.[29]

Dick's highly biased report for *Science* misrepresented the data as well as the arguments of the impact opponents. I took it personally. It was bad enough that he didn't fairly report on the science, but he was also actively trying to damage my career. I didn't care anymore who offended me or what bridges I burned. Dick had chosen to make me his public enemy, and I was going to hit back with everything I had. In reply to Dick's report (mid-February 1994), I wrote a letter to *Science* detailing thirteen of his most egregious factual errors and falsehoods. To increase the chance that this letter would be published, I sent it to every member of the editorial board and every trustee of *Science*. Norman MacLeod did the same, concentrating on the false blind test specifically, which of course I never saw because of the snowstorm on the last day of Snowbird III. Finally, after a long silence, *Science* published our two letters on April 29, 1994.

102 THE LAST EXTINCTION

The day after the letters appeared in *Science* magazine, Majda and I were sitting in the Small World Coffee having breakfast as usual, when Princeton University's President Bill Bowen appeared at our table and exuberantly congratulated me for having the courage to fight back against poor science reporting and to fight for truth in the impact controversy. He invited us to dinner at his residence, where we discovered him to be a charming and entertaining storyteller. Bowen's support cheered me up tremendously. It was just the lift I needed to continue my science war. There were many others who praised me for taking *Science* to task, including Dewey McLean, Charles Officer, Bill MacDonald, Bill Ward, and numerous academic friends and particularly mathematicians. It felt good. I had survived yet another kamikaze attack on my integrity.

Snowbird III was the turning point in my attitude toward the Dinosaur Wars. I was through dutifully collecting evidence, as if that was going to change the minds of the impact camp. It was futile. It took me nearly ten years to come to this realization. You'd think I would have reached this conclusion faster, but I was a stubborn idealist, and I persisted in believing I could appeal to their logic and intelligence, as well as our shared integrity as scientists seeking the truth. It took all the absurdity of Snowbird III to drive home the lesson that truth in science doesn't matter if scientists live in a self-isolating echo chamber.

So I gave up trying to convince the impact proponents of the contradictory evidence in the rocks right under their noses. I was through playing defense and refuting the evidence that the impactors produced. I abandoned diplomacy and let my new devil-may-care attitude fill my sails. I was going to take a new tack. I was going to go back out in the field and find the evidence myself. This shift in approach turned out to be the change that made all the difference.

8

CRAZY FUN
(SIX YEARS *of* DETECTIVE WORK)

AFTER THE DRAMA OF SNOWBIRD III AND ITS UNPLEASANT aftermath, I resolved to stop thinking so much about my enemies. Instead, I focused on my friends, my students, and the adventure of fieldwork. Wolfgang, Thierry, and I emerged from the conference with a lifelong bond. I finally had a team with whom I could enjoy fieldwork and in the coming years we co-lead many trips with my students.

Thierry, our sediment expert, was the pessimist. He met every new suggestion, interpretation, or even guess with an emphatic *no, no, no*. He could only think positively when the data was foolproof and assembled before him, and then he would run with it, sometimes claiming the idea as his own and denying ever having opposed it. Thierry called it the Swiss Technique: Be skeptical until the chickens come home to roost.

Wolfgang took the middle road. He was open-minded and based his interpretations on whatever evidence was available. He was always fun when discussing the pros and cons of our fieldwork as well as other life stories.

I was the daredevil optimist, knitting together evidence from different topics until the whole was larger than the sum of its parts. My mind's eagle eye saw how disparate topics related to each other and yielded more interconnected evidence. Consequently, I invited experts

104 THE LAST EXTINCTION

to join us whenever a new topic enriched our overall knowledge and could bring us closer to the truth.

The three of us were not only an excellent research team but fellow adventurers whose individual approaches to science and life were complementary. Our expeditions became famous at Princeton, not only for the learning experience they provided, but for the spectacular, exotic destinations we traveled to. Over the years, we journeyed to Mexico, Guatemala, Belize, Haiti, Cuba, Madagascar, Morocco, Tunisia, Egypt, Israel, Ecuador, Brazil, France, Spain, Denmark, China, and India. You can learn so much more in the field than you can in the classroom—and not only about science. The students who went on these trips were also exposed to new cultures and tested with the rigors of adventurous travel. As I knew from personal experience, when you take yourself to the remotest corners of the world, you also learn a lot about yourself.

If you want to peel back a person's layers and peek at their soul, take them traveling. If you want to expedite the process, take them on a geology field trip. Fieldwork has an uncanny way of sorting the worriers from the warriors, perhaps because it strips us of our comforts and lands us in unfamiliar places. Say you are doing research in the Sahara. You spend all day in the wilderness. Nothing is familiar. You don't speak the local language. Are there snakes? Scorpions? Better not reach under a rock or suffer the consequences of being bitten by worse than bugs. Meanwhile, the sun blazes down upon you, and sand dunes travel with the speed of the wind across the roads. When it rains there are few places to run for cover, and you may have to run to avoid sudden desert flash floods. Don't speed around curves if you are driving in mountains because the road may have washed away, leaving you and your vehicle in a free fall.

Every day, it's more than likely you've eaten something that doesn't sit well in your belly. There are no bathrooms, no trees, no rocks in sight. You just find a small depression to squat over, always keep your roll of toilet paper ready, and drink your bottle of Liquid I.V. for survival. All this so you can scrabble in the dirt all day. "Filthy" is a generous way of describing your appearance. Dirt has not only embedded itself beneath your fingernails, but it's also deep within your ears and nose. Dust and sweat encrust your face. Hot showers are not available; if you're lucky, a

CRAZY FUN (SIX YEARS OF DETECTIVE WORK) 105

splash of cold water is all you get at the best no-star hotel, and late at night you climb into bed between sheets that have seen multiple earlier visitors.

Under these conditions, you either shut down, dutifully do your work, and get out as soon as you can. Or you embrace the magnificent and rare opportunity life has given you to experience this place. So, you journey on camelback to camp out in the sand dunes with a blanket as your bed. You stare at the unfamiliar desert sky, mesmerized by the blinking of millions of stars. You feel the urge to get up and in total exhilaration tumble down the sand dunes screaming with happiness, only to wake your travel companions who join you in celebration. By the end of the field trip, you realize this is the happiest you've ever been in your life, and you gather your team for drinks and stories that last late into the night. When I run into former students today, they tell me how fondly they remember the field trips and my stories. A few have even likened me to a female Indiana Jones, and I confess I enjoy the comparison.

With Snowbird III behind us, Wolfgang, Thierry, and I leaped into a joyful period of collaborative work. Our mission: to pinpoint the precise age of the Chicxulub impact and uncover the real cause of the mass extinction. We had already pointed out Chicxulub's pre-KPB age and lack of tsunami evidence in our first investigations of northeastern Mexico localities at El Mimbral, La Lajilla, and El Peñón in 1993. To be darn sure we were on the right track, we began a comprehensive investigation of additional outcrops in northeastern Mexico with our undergraduate, master's, and PhD students.

Over the next eight years, from 1993 to 2000, our investigations revealed dozens of impact spherule deposits spread over approximately 350 kilometers across northeastern and southwestern Mexico. This area has low mountains from 500 to 1,000 meters high with an abundance of impact glass spherules reworked and redistributed in latest Maastrichtian sediments prior to the KPB mass extinction. Finding the *primary* impact spherule layer remained elusive.

Under normal circumstances, the evidence we already collected should have been enough to put the impact theory to rest. But the impact controversy was everything but normal. We needed an overabundance of proof to demonstrate the pre-KPB age of the Chicxulub impact. It

106 THE LAST EXTINCTION

felt like a blissful return to pure science. We were no longer chasing the impact group, working to confirm or disprove their conclusions. We were independent seekers of new information. I felt as though I'd tossed a hundred-pound weight off my back.

By 1999, a very promising impact locality was Loma Cerca B, located on a steep hillside fifty kilometers to the north of El Peñón. There, Thierry's master's students discovered a *primary impact glass spherule* deposit.

The Loma Cerca B outcrop consists of a 3.5-meter-thick sandstone that spans the top of the hill and lacks fossils. Below the sandstone are 3.6 meters to 17 meters thick marlstones interrupted by two separate impact spherule layers. The upper spherule layer is one-meter-thick (9.9–10.9) and consists of impact spherules mixed with marlstones and current activity. The lower impact spherule layer is fifty centimeters thick and consists of undisturbed impact glass spherules.

This was huge, like finding a rare giant Galapagos tortoise. The implications of this discovery took several months to a year to sort out what really happened. *Primary impact glass spherules* are the first original spherule fallout that rained from the sky and settled on the deep seafloor in quiet deep water, where it remained undisturbed for hundreds of thousands of years. We had discovered the first *primary* impact glass spherules.

The giant Galapagos tortoise metaphor, or analogy, has implications for protected species like desert tortoises, as well as the *primary* impact spherule deposits which are rarely found. But what about the common small desert tortoises often kept as pets, like my Footsy, Stinky, and Icarus, that can be analogous to common *reworked* impact spherules. These are the *"secondary"* impact spherule deposits which have been mixed, reworked, and eroded from shorelines and carried away by active currents in shallow waters. Geologists call such deposits *"secondary"* or reworked deposits because they lack *"primary"* deposition. *Secondary* reworked impact spherules are common to abundant throughout northeastern Mexico, Belize, Haiti, Cuba, Texas, and the northeastern US. But they are not useful for dating reworked sediments of unknown age. For this reason, we searched for *primary* impact spherule deposits for many years.

The lower spherule layer at Loma Cerca B is fifty centimeters thick (12.2–13.3m) and consists of pristine impact glass spherules.

This spherule layer descended through quiet, deep water and landed on the seafloor undisturbed at about five hundred meters depth for hundreds of thousands of years. These are the deep-water *primary* spherule deposits that remained pristine and unaffected by currents and tectonic activity.

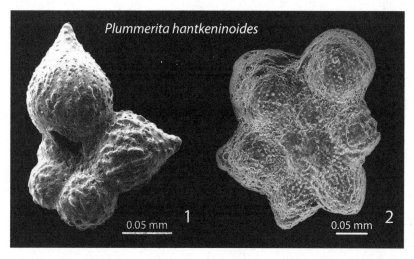

Plumerita hantkeninoides *was the last species that evolved about 220,000 years prior to the KPB mass extinction and went extinct at the mass extinction. This species is the most critical age marker and recognized by its small to medium size, four to six spherical chambers, surface ornamentation, and exotic elongated spines.*

Now that we know how the *primary* spherule layer was deposited, we investigated the depositional age of this impact. Dating is based on characteristic foram species with short or overlapping life spans that yield good age control. There is just one foram species, *Plummerita hantkeninoides*, that evolved approximately 220,000 years prior to the KPB mass extinction and went extinct at the KPB. This species is easily recognized by its small to medium size, four to six spherical chambers, rugose surface ornamentation, and exotic elongated spines. The time in which this species lived is expressed over the corresponding area of sediment in which it is found, known as Cretaceous foram zone CF1.

108 THE LAST EXTINCTION

At the base of zone CF1, or 220,000 years pre-KPB, we saw marl-stones deposited along with abundant forams. Nothing unusual about that. But then at about 200,000 years impact glass spherules settled on the seafloor. This confirmed the age of the Chicxulub impact predated the KPB mass extinction by approximately 200,000 years, based on the impact evidence's location in the early part of foram zone CF1. We con-cluded this was the *primary* spherule fallout that could only have come from the Chicxulub impact. This, again, was more evidence suggesting that Chicxulub's impact was relatively minor on the global scale, causing neither significant extinctions nor long-term climate change. In other words, Chicxulub's impact and environmental effects had been vastly overestimated.[30]

We presented this data at Snowbird IV in 2001. Though it was yet another piece of positive evidence dating the impact long before the fifth extinction, impactors again conveniently explained it away as the product of the tsunami earthquake disturbance, for which there was no evidence. Still, I had to concede that our evidence of Chicxulub's pre-KPB age needed to be stronger. Now, it rested only on one outcrop of *primary* impact spherules dated about 200,000 years pre-KPB. This was not convincing enough to unseat the belief that the Chicxulub impact predated the KPB. To make our case stronger, we needed at least one more locality with a primary impact spherule layer.

In March 2000, on a field trip in Mexico, a particularly eager group of my students launched their own investigation to find the primary impact spherule fallout from the Chicxulub impact. I was pleased with their passion, but also apprehensive. I did not want to see their enthusiasm for science squashed when their efforts to find this needle in a haystack failed—as it almost surely would. Nevertheless, I suggested the best bet was the El Peñón hillside, which we had earlier investigated and found to contain abundant *reworked* impact spherules. We could search for an older outcrop that had never been investigated. Could this reveal the *primary* impact spherule deposit? The chance was miniscule.

The next day was a ghastly 115°F. Wolfgang and I loaded up our vehicle with water bottles, oranges and bananas, picks, and shovels, and we drove the students to El Peñón's highest hillside, which we had never investigated before.

"Start digging a deep trench as far downhill as you can from that so-called 'impact tsunami' sandstone near the top," I advised. "If you're very lucky, you might find the primary impact spherule layer, but don't count on it."

In the blazing heat, three students swung their picks and heaved their shovels into the dry earth to dig a one-meter-deep trench through the undisturbed horizontally layered marls, which are calcium carbonate–rich mudstones of the former seafloor. It was brutal labor in the heat and still they kept digging.

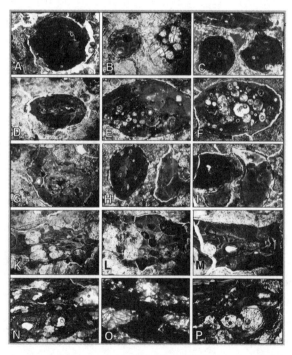

El Peñón cooling history of primary impact spherules (about 200,000 years pre-KPB) from bottom to top: (1) At the base, hot glass rained from the sky and welded into ten-centimeter-thick rafts on the sea surface which rapidly sank to the bottom at about five hundred meters; (2) subsequent rainout of hot glass retained variable spherule shapes: spherical, oval, compressed, convex, concave contacts; (3) Chicxulub impact glass from the two-meter-thick primary spherule layer at El Peñón ended with cooled, smaller impact glass fallout; (4) calcite precipitation settled between impact spherules on the sea floor.

110 THE LAST EXTINCTION

That afternoon, the trench diggers hit the jackpot. They uncovered a two-meter-thick impact spherule layer just four meters downhill where the trench revealed a dark, glittering layer of impact glass spherules devoid of shallow water debris and sediments. Never before had a pure impact glass deposit of such concentration been encountered. In the four meters, higher up normal marls revealed no sediment mixing, nor did the one meter below the base of the two-meter-thick impact spherules. Rich assemblages of planktic forams showed an older age zone CF1, dated approximately 200,000 years below the impact spherule layer, the same identified the year earlier at Loma Cerca.

For the second time we'd uncovered the primary spherule fallout from the Chicxulub impact's mushroom cloud—and it was even more convincing than the first. The two-meter-thick concentration of impact spherules—ten times the amount at Loma Cerca—erased any lingering doubts of Chicxulub's pre-KPB age.

We contemplated this new discovery in stunned silence, not daring to believe our luck. The bold move by my students and my educated guess had paid off. My students grew euphoric, whooping and hollering, laughing, and chattering away as they used their rock hammers to cut and grab chunks of the spherule rock as mementos of a history-making discovery.

Once the students calmed down, we reflected on what had happened. Here, in this dry desert rock that was once a seabed, recorded the immediate aftermath of a truly spectacular event. After the Chicxulub asteroid hit, impact glass rained from the mushroom cloud of dust and melt rock generated by the impact. When the first hot melt rock glass accumulated on the sea surface, it melted together and formed rafts of welded glass up to ten centimeters thick that rapidly settled to the seafloor. As this impact glass cooled, larger oval and compressed spherules first settled on the seafloor and stacked up followed by smaller ones. This was the primary fallout from the impact because it consisted only of melt rock and impact spherules directly deposited from the fallout within minutes.

When you look at the top few centimeters of the two-meter-thick spherule deposit, you see smaller and dispersed spherules marking the last remnants as the fallout deposit. Carbonate sedimentation resumed quietly. There was no current activity in this deep water of

CRAZY FUN (SIX YEARS OF DETECTIVE WORK) 111

five-hundred-meter depth. The two-meter-thick impact deposit quietly ended. Sedimentation resumed and planktic forams survived the impact, though less abundant than before. There is no sign of current erosion and redeposition. Just quiet deep-water sedimentation.

Having wrapped up a wildly successful field trip, it was time to take El Peñón samples back to the lab for analysis. From them we could get more data about the age of the Chicxulub impact and how the planktic forams responded. Our foram analysis yielded nearly identical results to Loma Cerca but lacked the initial melt rock and early sequential size difference.[31] It showed the primary spherule fallout dated approximately 200,000 thousand years before the KPB mass extinction sixty-six million years ago. We quantified the response to the impact based on the relative abundance of each species population across the primary spherule deposit and attributed any significant changes to the impact. There were none. No species extinctions and no significant environmental changes.

The evidence was clear. The Chicxulub impact crashed into Yucatan ~200,000 years prior to the mass extinction and left no long-term effect on planktic foram populations, which went extinct at KPB time. This was a tremendous revelation. A large impact leaving a 175-kilometer-in-diameter crater and an impressive layer of glass spherules had no significant impact on foram populations.

It was nearly a decade before we could publish our findings. Our progress was initially delayed due to the drug war in Mexico, which made it impossible to return to El Peñón to collect more samples until 2007. Even then it was a dangerous trip for Thierry and me as no Mexican colleagues dared to venture into the drag war. From 2007 to 2009 the delay was due to the machinations of the impact camp. Collaborators Jan Smit and Peter Schulte repeatedly selected each other separately as peer reviewers for our paper and each rejected our study three times in separate journals. After each rejected review, I wrote to the journals to protest the bias of their reviews. Finally, in 2009, the London Geological Society editor accepted our paper and published it rapidly,[32] but impactors continued to ignore our data.

Over the next few years, we scoured areas around the Caribbean for more impact spherule layers. In the summer of 2000, we went to Haiti.

112 THE LAST EXTINCTION

This was now nine years after impact spherules were first discovered in a roadcut near Beloc, Haiti, triggering the race to discover the impact crater. We had arrived to reinvestigate this evidence that claimed to prove the Chicxulub impact was KPB age.

Near the city of Beloc, a roadcut revealed Cretaceous impact spherules mixed with Paleocene rocks, which reflect tectonic activity—not surprising on an island that frequently suffers from violent earthquakes. The next outcrops revealed impact spherules scrambled across twenty-five centimeters of mixed Cretaceous limestone and early Paleocene forams. Such mixed deposits mark redeposition well *after* the mass extinction and reworked sediments. The spherules and underlying Cretaceous had been eroded and scoured by currents. All that was left were residual spherule deposits in the overlying Paleocene sediments. Everything was scrambled, mixing layers of earth that were a couple of hundred thousand years apart. In Haiti, impactors failed to recognize this mixed bag with uncertain age and origin. After our publication on Haiti, impactors stopped talking about the Haiti sections; it was clear the science was all wrong.

Later that year, Wolfgang and I ventured to Cuba, where impactors claimed the asteroid impact on Yucatan sent rocks the size of cars flying hundreds of miles through the sky to land in Cuba and other Caribbean islands. This implausible assertion erupted in heated critique from geologists who knew a great deal about Caribbean geology and how the islands became what they are today. That killed off the argument but not the impactors' belief.

In Cuba, we traveled through the country in a government car with a woman professor devoted to Fidel Castro. Running out of gasoline in Santa Clara, the driver took a can and went on foot to search for fuel while we waited in the car. Hungry, I unpacked a four-kilogram wheel of gouda cheese to snack on. This was the only food I had managed to bring through airport security. I started cutting pieces with my Swiss army knife and in no time dozens of children surrounded the car asking for food. I cut slices of cheese as fast as I could, and Wolfgang passed them through a slit in the window—the cheese wheel disappeared, leaving still dozens of hungry children. The professor tried in vain to shoo them away, saying repeatedly, "This is not Cuba, people don't go

CRAZY FUN (SIX YEARS OF DETECTIVE WORK) 113

hungry here, all have enough to eat." We knew better. That morning, we observed people waiting in long lines at food stores that were nearly empty.

That trip to Cuba deepened the mystery of why we were finding impact spherule layers at so many different depths throughout the Caribbean. Only in Santa Clara did we find small pockets of impact spherules about two centimeters thick in mudstone. As in Haiti, deposition occurred in the early Paleocene. Something was going on in the Caribbean that eliminated impact spherules in Cretaceous sediments and dumped them in the early Paleocene—just like in Haiti.

Next, our team struck out to visit Belize and Guatemala. With each destination, our fieldwork experiences grew wilder and weirder. As with Cuba, impactors claimed rocks the size of a car or even small house were said to have hit Belize, supposedly within minutes after the impact crashed into Yucatan. One of those stories claimed that giant boulders were thrown five hundred kilometers into a quarry near the Belize/Yucatan border. It sounded like another piece of science fiction, but we went to check it out.

Punta Gorda, affectionately called PG by locals, is located at the southern tip of Belize where the road ends in jungle. It seemed like you could see the end of the world. PG was barely accessible by car on an unpaved potholed road at no more than twenty-five kilometers per hour. The better transportation was a "puddle jumper," a four- to six-seat plane that could land at Punta Gorda's airport, which was little more than a Home Depot garden shed painted in vivid purple/yellow/red that also served as a beacon guiding the arriving plane. Brian Holland awaited us at the PG airport on our trips.

Brian Holland is a geologist and expatriate American who introduced us to many of the joys and pleasures of Belize. This included Mayan ruins, jaguars, wild pigs, crocodiles, a two-meter-tall stork, and the country's fabulous reefs. Some of the expatriate Americans we met were equally exotic and weird, living in this jungle paradise in "homes" with no walls or roof, their bedrooms surrounded by exotic tall flowers. They would sleep unperturbed by the nightly visits of a jaguar skirting their bed eyeing their next meal.

114 **THE LAST EXTINCTION**

Brian operated a dolomite quarry of magnesium-rich limestone used as fertilizer in agriculture, which was funded by the World Bank and was the only employment in Punta Gorda. He was an interesting character. Middle-sized, wiry, and energetic with blond hair, crinkly eyes, and a ready smile that might burst into full-throated laughter at any time. He was the type of person you took an instant liking to. He was a beloved sort of Pater Noster to his employees, and a last resort for the town's people as he took care of just about anyone from Mayans and Garifuna to the expatriates who wound up living in PG.

Our team searched throughout Belize, from the Albion Quarry in the northeast to PG in the south, for primary spherules deposits. But just as we observed in Haiti and Cuba, thick deposits of impact spherules were invariably found in early Paleocene sediments.

In 2000, thanks to a deep drilling operation in Belize, I had an excellent chance to find the primary spherule fallout and test the impact crater's age. The drilling rig was offshore from the Monkey River, not far from Punta Gorda. The owner of this drilling operation was Steve Reilly, a unique, globe-trotting American who used his own ship and machinery to prospect for oil with the hopes to make millions. As in any drilling operation, like the Japan Trench when I was still a graduate student, most critical is drilling into successively older age sediments. As a paleontologist, it's easy to find which direction is older or younger, but if you lose that, it's over. Steve offered me the cores he drilled in return for my age dating and I readily agreed. Alas, within weeks, a hurricane capsized the rig and platform while it was being towed to the shore to escape a hurricane. The cores were lost.

Our last resort was a couple of outcrops located in Guatemala, just fifty kilometers to the west of PG as the crow flies. Though it was close, getting there would be a challenge. PG might as well have been the end of the world. It had no roads leading anywhere. Add that to the fact that eastern Guatemala was still embroiled in its tragic thirty-year civil war. Hiring a puddle jumper to fly us over this war zone would be insanity. Therefore, only the sea was an option to get to Rio Dulce on the Guatemala side. Brian Holland arranged for a ride on a small fishing boat that made weekly runs to Rio Dulce. Majda had agreed

CRAZY FUN (SIX YEARS OF DETECTIVE WORK) 115

to join me on this trip to Belize. While I was doing fieldwork, he had been holed up alone at the Smith Hotel in PG, and though he rarely wanted to leave town, there is only so much a man can take alone at the end of the world. He decided to join Wolfgang and me on the trip to Guatemala.

Once on board the boat, we asked Brian why it made weekly runs to Rio Dulce. Then, with more urgency, we asked why the boat was leaking. That's when we noticed the many bullet holes in the boat's hull. Terror gripped us. We realized we had unwittingly booked passage on a drug smuggling boat to Rio Dulce. The thought was beyond scary. We pressured Brian to fess up. He was unusually nervous and scared but confessed. It was a weekly drug boat run that had been involved in a shootout with a rival gang the previous week, leaving some dead and injured, including the owner of the Smith Hotel. On this trip, the Belize drug smugglers feared there would be another shootout before we reached Rio Dulce. We all went mute as the fear gripped us. We kept our eyes glued to the horizon for the rival drug gang and prepared for the worst. An assortment of guns was pulled out by the Belize drug gang and handed to us to prepare for the coming battle. Fortunately, we reached Rio Dulce without incident and hightailed it out of town as fast as we could to a hotel located a reasonably safe ten kilometers out of town. It was a dreadful night full of nightmares.

As if the trip wasn't enough, the next morning began with continuous monsoon-like rain. After, Brian left to take the drug boat back to PG. Majda stayed in the hotel room fearful the drug gang would find us. Wolfgang and I went in search of the Actela outcrop in the heavy downpour. We collected impact spherules along the rapidly rising Actela River. By late afternoon, we relocated to a safer hotel within Majda's comfort zone.

The next day a creaky old American school bus painted in bright cheerful scenes took us back to Guatemala City. Wolfgang returned home. Majda and I went to Antigua, the beautiful Spanish colonial city where we relaxed at the Quinta de las Flores Hotel and watched the spectacular nightly volcanic eruptions from the safe distance of our hotel garden. The traumatic drug boat ride receded into memory. We fell in love with the western Guatemala highlands and the Mayan people, and Guatemala soon became our regular vacation destination for many years.

By 2006, it was time to take stock of all the research Wolfgang, Thierry, and I had done. Eight years of detective fieldwork around the Chicxulub crater in Yucatan and in other areas from Mexico, Guatemala, Belize, Cuba, Haiti, North America, Caribbean (Gorgonilla), and other localities had ended with great success. We had analyzed impact spherules of varying sample sizes collected from the Caribbean through Mexico. My KPB archive at Princeton overflowed and became the largest collection of impact spherule samples and the KPB mass extinction in the world.

But what did the pieces add up to? After all those years of work, it was time to assemble the puzzle, step back, and look at what really happened sixty-six million years ago, supposedly when the Chicxulub crater formed.

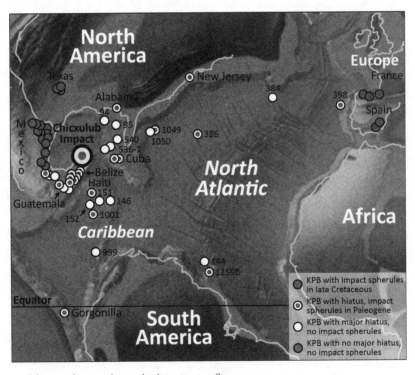

Paleogeographic map of impact localities sixty-six million years ago.

CRAZY FUN (SIX YEARS OF DETECTIVE WORK) 117

If we could travel back in time sixty-six million years, the world would still be recognizable to us but look much different. There were no polar ice sheets. Consequently, the sea level was two hundred meters higher, inundating lowland areas from New Jersey down to Florida, as well as the Yucatan, Belize, Guatemala, and parts of northeastern Mexico. North America was split in two parts by a vast Western Interior Seaway that stretched from the Gulf of Mexico to the Arctic Ocean. The Caribbean was a vast seaway connecting the Pacific and Atlantic Oceans.

Onto a map of this ancient Earth we plotted the ninety-three localities we analyzed, many of them with multiple KPB mass extinction sequences. In fifteen of these localities, the time window of the spherule fallout and redeposition was missing (white circles). Of the seventy-four remaining localities, fifty-two are from northeastern Mexico and Texas with impact spherules in the late Cretaceous below the KPB mass extinction (filled-in circles). Twenty-two impact spherule localities are from south and southeastern Mexico up to the North Atlantic as far as New Jersey (gray dots in white circles); all of these are reworked impact spherules from the early Paleocene. In addition, a scattered small number of reworked impact spherules are found in early Paleocene sediments from the northeastern Atlantic Ocean. In these late Paleocene localities, impactors consistently claimed the age was precisely the KPB mass extinction. They consistently ignored the truth of the pre-KPB age because it didn't fit the mass extinction.

What emerges from this picture are two distinct age and spherule distributions patterns. These patterns, and the different interpretations of them, are at the root of the acrimonious, four decades–long Dinosaur Wars. Pattern #1, which is restricted to northeastern Mexico and Texas, shows all impact spherules *were of Cretaceous age below the KPB mass extinction*. Pattern #2 covered the area where erosion and spherule redeposition *always occurred in the early Paleocene well after the KPB mass extinction*. Neither pattern confirms the impact theory, but they do pose a mystery. Why is the impact appearing at different times in the geological record?

Our analysis revealed something new and amazing. We discovered that the spherules were mixed up because, for tens of millions of years, a powerful ocean current that caused major sediment erosion had

118 THE LAST EXTINCTION

continuously carried away sediments and redistributed them through the Gulf of Mexico and up through the North Atlantic. And it wasn't just any ocean current. The spherule redistribution pattern was so clear that the answer stared me in the face: We had mapped the erosion pattern of the Gulf Stream, which I had first documented in the Caribbean in 1993.

The Gulf Stream has been active since the Cretaceous, moving north from the Antarctic, through the South Atlantic, veering west into the Caribbean and eastern Gulf of Mexico before turning east into the North Atlantic along the US coast. It never reached into the northwestern Gulf of Mexico, which is the reason there is no major gap in the sediment record of northeastern Mexico or Texas. The absence of this current movement is why the primary impact spherule layer was preserved, and why the subsequent shallow nearshore erosion and redeposition was preserved. The answer had always been there but because everyone was focusing on the impact theory and tsunami, no one was thinking outside the box.

Overjoyed by our results, we wrote two papers: one for *Earth Science Review* titled, provocatively and tongue-in-cheek, "Multiple Impacts Across the Cretaceous-Tertiary Boundary." The title was a nod to the impactors' love of comet showers and a refutation of their claim that impact spherules in deep-sea cores were not the ultimate proof of Chicxulub's KPB age. We linked the multiple spherule erosion and redeposition pattern during the early Paleocene to global cooling, which intensified the Gulf Stream current and caused repeated erosion. The cause for this distribution pattern was global cooling in the early Paleocene and Gulf Stream erosion. And for the first time, we unequivocally linked Deccan volcanism to global climate change.

Our papers received very good peer reviews. Strangely, there were no negative reviews from impactors. This was surprising—usually we could expect strong pre-publication criticism. But somehow that disappeared from the impactors' radar. Was it too complex? We had been digging for evidence in areas they felt had already proved the KPB impact theory, and they had bigger fish to fry. And so, they had focused their energy and attention on the coming Chicxulub drilling, which they believed would prove once and for all that the impact caused the KPB mass extinction. And that's where they went wrong.

9

The MYSTERIOUS CASE *of the* VANISHING CORES

IN THE FALL OF 2000, WHILE I WAS ENJOYING OUR DISCOVERY of the primary impact spherules at El Peñón and Loma Cerca, the National Science Foundation's Continental Drilling Program approved funding for scientific drilling through the Chicxulub crater. The primary objective of the drilling was to prove that the impact that created the crater happened at the time of KPB mass extinction and was therefore its cause. The announcement of the program was major news. I welcomed the effort. It meant no more scrambling to get my hands on incomplete core samples from PEMEX or dealing with dissembling impact scientists unwilling to share. For the first time, the critical core samples of layers before and after the mass extinction would be available to all scientists who submitted a proposal and received approval. At least that was how it was supposed to work.

I could see the NSF's plan for the drilling and the distribution of samples was compromised from the start. Scientific crews on drilling expeditions are usually diverse and inclusive, but the Chicxulub drilling team was the opposite. And though the NSF required that access to the drilling samples be based on scientific merits, regardless of differing interpretations or beliefs, the Chicxulub drilling team ignored NSF rules and limited participation and access to the core samples they obtained to true believers.

120 THE LAST EXTINCTION

In January of 2001, the principal investigators (PIs) of the drilling research project announced that scientists interested in participating in the crater core research must submit applications for approval. The PIs would compile a list of scientists approved for participation and submit it to NSF's International Continental Scientific Drilling Program (ICDP), which would consider funding proposals only from this prescreened list. Would-be participants thus had to pass two hurdles: the drilling team's prescreening test, followed by the standard peer review system. Because there were impact supporters in each of the groups that screened applications, the chances of success were remote for anyone but the Chicxulub drilling in-group.

In the face of these daunting odds, I immediately submitted my proposal to study the core across the KPB to the impact drilling team with copies sent to three individual PI members as backing. The PIs approved my proposal, and my name was added to the list. I believed I'd passed the first hurdle, but then my name mysteriously disappeared from the approved list. As it turned out, the list was managed by Smit, which explained why my proposal was the only one eliminated of thirty-six submitted. It was only after I tracked what happened to my proposal after the three PIs had approved it that Smit admitted he single-handedly eliminated my proposal. After Smit's action, my name was restored to the proposal list.

By this point, I realized impartiality was out the window and that there would be little hope of getting funding from the NSF, because biased peer reviews from impact proponents would torpedo any dissenters, especially if the dissenter's name was Keller.

At the end of March, I received a phone call from Leonard Johnson, NSF's director of continental drilling (ICDP), whom I had never met. He asked me to provide a list of potential reviewers.

I laughed out loud, saying, "It depends on whether you want to fund the proposal or not. There is no one on the impact side that wants me to work on the Chicxulub cores. So, if you want to reject the proposal, just pick anyone on that side of the controversy. But if you are interested in a more balanced assessment of my work, take any knowledgeable scientist not rooted in the impact camp."

THE MYSTERIOUS CASE OF THE VANISHING CORES 121

This group included the well-known anti-impactors and pro-volcanologists Dewey McLean, Charles Officer, and their collaborators. But it also included the silent majority who had been hounded out of KPB work and silenced by the impact group more than ten to fifteen years ago. I didn't provide Leonard with their names because no one wanted to be known publicly for fear of reprisals, nor did they ever want to be acknowledged as reviewers of our publications for the same reason.

Later that summer, I was certain my proposal would be blocked—and why not after all the shenanigans I've already witnessed? But then I got a call from Leonard Johnson that my proposal was approved with full funding. It was a relief to know that at least some ICDP members were still committed to a scientific process.

This began over a decade of funding by ICDP Director Leonard Johnson and the Sedimentology and Paleontology Program Director Rich Lane. They believed my small team should be heard. It was a welcome example of scientific skepticism being encouraged.

From December 2001 to February 2002, the Chicxulub Scientific Drilling Project (CSDP) gathered samples at a site near Hacienda Yaxcopoil, a former colonial plantation located south of the city of Merida in the Yucatan. In the spring of 2002, the select group of scientists who had submitted the thirty-six proposals approved by the PIs for studying the Chicxulub cores, were invited for a one-day "sampling party" at the National Autonomous University of Mexico (UNAM) in Mexico City. One representative from each team would be allowed to examine the Yaxcopoil-1 cores (Yax-1 for short) and submit a list of samples they wished to study. Why they called it "sampling party" was unfathomable because nobody could even touch the cores, let alone obtain samples.

The "sampling party" was managed by the four Chicxulub drilling PIs: Jaime Urrutia-Fucugauchi, who housed and controlled the Yax-1 cores at UNAM; Virgil (Buck) Sharpton from NASA; Jan Smit from Amsterdam; and Dieter Stöffler from Potsdam, Germany. The three dozen proposal representatives gathered in a large room where the Yax-1 samples were laid out in standard white plastic core boxes, each containing four core segments eight centimeters wide and one and a half meters long. A meter

122 THE LAST EXTINCTION

and centimeter scale set alongside the core boxes revealed the depths of the core intervals.

Wolfgang Stinnesbeck was my team's representative at UNAM's "sampling party" because he lived and worked in Mexico and had friendly contacts with their team. We kept in close contact by phone and email through the day to keep abreast of the sampling progress and prevent further shenanigans we expected at any time.

Two months of drilling between depths of 400 to 1,511 meters below the seafloor revealed the drilling team's most critical core intervals. The first discovery was just above a K–P hiatus and revealed the incomplete KPB boundary event. The second discovery was a fifty-centimeter-thick fine-grained Cretaceous limestone well below the KPB. The third discovery was the one-hundred-meter-thick impact breccia—the result of the Chicxulub crater impact.

The excitement was palpable as scientists examined the cores they wanted to study. There was a steady murmur in the room, occasionally pierced by shouts of excitement. As the representative of our team, Wolfgang searched for the samples relevant to our two proposals, which included mine to study the critical fifty-centimeter-long core, and his to study for the entire 736 meters of core below the KPB. Thanks to his height, he was able to get a bird's-eye view of samples on the tables throughout the room and was quickly able to find the ones we wanted.

Soon, everyone homed in on the table displaying the critical KPB intervals. Scientists salivated over this section of the core because it was unarguably the most important piece of Chicxulub impact coring and the primary reason for drilling the impact crater in the first place. This core could settle the age of the Chicxulub impact for good. It potentially held the proof—or disproof—that the Chicxulub impact triggered the mass extinction.

The PIs instructed the "sampling party" to paste small Post-it labels marking the portion of the core they wanted to study. We wanted to be thorough and requested samples at five-centimeter intervals, which was the most anyone could officially obtain. From the uppermost part of the impact breccia to the dark gray-green KPB layer, the Post-it labels of

THE MYSTERIOUS CASE OF THE VANISHING CORES 123

multiple teams quickly covered up the core, accumulating in a thick wad of multiple layers that soon lost all usefulness. Eventually, the Post-its were removed and sample requests were recorded on a sheet of paper. Jaime promised to ship samples by the end of May.

And then, poof. The most critical fifty-centimeter KPB core vanished. Someone had physically removed the most coveted sample from its table.

Chaos broke out. Everyone became a suspect. Jaime locked all the doors to prevent anyone from leaving as the search progressed frantically. The other co-PIs, Buck Sharpton, Jan Smit, and Dieter Stöffler, helped in the search. Wolfgang relayed updates to us over the phone. I wish I could say I was surprised, but this was the *third* time that the most critical KPB core interval from the Chicxulub crater had vanished. The first time was when the PEMEX cores were supposedly burned in a warehouse fire in Mexico. I suspected the warehouse fire was invented to prevent the study of these cores because PEMEX consultants in the 1970s had already identified Cretaceous limestones above the impact breccia, which would mean the Chicxulub impact predated the fifth mass extinction and could not be its cause.

It wasn't until 1994 at the Chicxulub drilling workshop that Buck Sharpton announced that the PEMEX core had miraculously reappeared, and that there never was a warehouse fire in the first place. The reason for this announcement, you may recall, was the discovery of the split core halves at the University of New Orleans in Bill Ward's lab where he inherited them from his predecessor, a PEMEX consultant. When Bill first heard me, the PEMEX core halves were in his possession, and he was asked to return them to Mexico immediately. But Bill and I agreed to collaborate on the study of those cores prior to their return. Within a year (in 1995), we confirmed the presence of Cretaceous limestone above the impact breccia. By then, however, the presumed KPB age of the Chicxulub impact put forth by the impactors had become widely accepted and no amount of hard evidence could shake that belief.

The second time the critical KPB core between the impact breccia and early Paleocene vanished was shortly after the 1995 drilling of the impact crater by UNAM. Jaime oversaw the cores and permitted Wolfgang to

124 THE LAST EXTINCTION

sample them above and below the critical KPB core interval, promising we would get samples of that core later. Wolfgang and I analyzed the core samples that were made available to us but never received the critical samples that showed the KPB boundary, despite repeated calls. Wolfgang even visited Jaime at UNAM, where Jaime confessed he was distraught because that critical core segment had mysteriously disappeared—taken, presumably, by someone at his own institution. Of course, nobody fessed up. We had all the data, except for the critical interval, and so nobody could ever publish the incomplete results.

And now, for the third time, the most critical core had vanished. It was like a disappearing act. But whoever this Houdini was, he wasn't interested in entertainment. He wanted to subvert the discovery of truth. What made this crime worse was the fact that the perpetrator was surely one of the over three dozen scientists at the sampling party.

An awkward silence engulfed the room as Jaime questioned the participants for clues about who took the core. And like an Agatha Christie novel, nobody fessed up. How was this possible? How could someone abscond with the most critical core in this crowd of scientists without being seen? It wasn't like you could hide a fifty-centimeter-long core sample in your pocket and walk away. It would stick out by nearly two feet. Were there accomplices? Who were they?

Based on my experience, I had my suspicions. The person who got away with it "unnoticed" was likely shielded by accomplices. They had to be someone with purpose and clout, who would feel confident they could brazenly steal the sample and get away with it. I imagined it was one of the Chicxulub drilling PIs. But which one? I knew it wasn't Jaime—he was obviously frantic about the core disappearance. It wasn't Buck's style to clumsily kidnap the core. That left just two suspects: the German PI, Dieter Stöffler, and Jan Smit. These two were like two peas in a pod.

While most sample party members were now uncomfortably standing around waiting, Jaime and his helpers continued the frantic search, which soon progressed to searching personal luggage. Not long after, the KPB core was discovered in Jan Smit's duffel bag. As if the theft wasn't galling enough, the entire time the search was conducted, Smit had cunningly pretended to help find the missing core to deflect suspicion. Once the

THE MYSTERIOUS CASE OF THE VANISHING CORES 125

ruse was up and he was caught red-handed, Smit said he was asserting his right to take the core to Amsterdam, where he could do a better job than the Mexican scientists of distributing the samples to all the interested parties. He guaranteed he could get it done by June. No one, of course, had authorized Smit's undercover kidnapping of the core and his offer to be custodian of the samples was rejected. Smit returned to Amsterdam in a huff and Jaime, who was outraged by the whole affair, restored the core to UNAM's collection.

If only the story ended there. A few weeks later, Smit flew all the way back to UNAM and, somehow, claimed the KPB core with the permission of all the other PIs, except Jaime. It's not clear how and why this happened, but I suspected that it was motivated, again, by impactor insecurity about what the hard evidence from the crater might reveal and a strong desire to control the narrative. I began to wonder if Smit and Dieter had planned to secretly eliminate the Yax-1 KPB core at the sample party out of fear that the evidence supported a pre-KPB age for the impact crater. Why else was this sample alone selected for special treatment? The previous mishandling of the PEMEX and UNAM cores had never been punished, so they had every reason to believe they could get away with it again. Except this time, Smit and his accomplices had been caught simply letting the KPB core vanish. This was no longer an option, at least not in the immediate future. The cat was out of the bag. Everyone knew what they were up to.

After the theft was exposed, they had to alter their plan and get the US PIs on board, leading to a truly disgraceful strategy. Smit had already started it the minute he was caught at UNAM: *The Mexicans can't be trusted to supply the samples to the researchers in time. Smit can do the best job out of Amsterdam.* This enormous insult to the Mexican PI and his team would come back to haunt them. But to the prejudiced US and European scientists, Smit's claim rang true and half of them believed Smit would keep his word. But Smit never intended to keep his word—not in the slightest.

10

RACE *to the* DEADLINE

GEOLOGISTS ARE ACCUSTOMED TO DEALING WITH VAST amounts of time. We can see in a cliff face the history of thousands, tens of thousands, even millions of years. With little more than rocks and fossils as clues, we reconstruct the story of what happened over these vast periods. Our scientific work, under the best of circumstances, needs to be slow, meticulous, and deliberate. I am not accustomed to feeling rushed, and yet, in 2002, with an important deadline approaching, my colleagues and I found ourselves measuring time not in eons, but in weeks, and even days.

That deadline was April 2003, when the first joint European Geosciences Union (EGU) and the American Geophysical Union (AGU) held their Grand Conference in Nice, France. These meetings are dedicated to the pursuit of excellence in the Earth and Planetary Sciences. Never had these two prestigious earth and space sciences organizations shared a conference. Everyone knew this was going to be a major event, attracting international media attention and drawing an attendance of over 10,000 American and European scientists—an incredible opportunity to present evidence to a large, diverse group of scientists. Finally, on this large, international stage, both impactors and skeptics would present the culmination of decades of competing claims about what really happened sixty-six million years ago. What's more, these arguments

128 THE LAST EXTINCTION

would be buttressed—or torn down—by the evidence unearthed by the Chicxulub crater drilling project.

We knew that in Nice the impactors planned to unveil, with great fanfare, their evidence purporting to show that the Chicxulub impact was KPB age. And we also knew that if we couldn't contribute our own analysis by the deadline, their unchallenged assumptions about the impact theory would be established as proven, and they would triumph. My colleagues and I had accumulated a mountain of evidence over fifteen years of hard work in the field and in the laboratory that consistently contradicted the impact theory. But this evidence was routinely denied, ignored, and even suppressed by the scientific establishment. But this time, we would have the same crater core samples to study and an open venue in which to share our results. Denial of our evidence in front of an international group of scientists was no longer a surefire option to win. Ultimately, even long-denied truth would prevail.

The first scientific drilling into the Chicxulub impact crater on Yucatan had been completed in January 2002. This meant the impactors who controlled the samples would have over a year to prepare for their presentation. They did not waste this extra time. Even during the two-months-long drilling process, their scientific team, consisting only of the top impact group members from various institutions, had been priming the publicity pump, reporting daily progress on their Chicxulub impact drilling website. Preemptively, the website reported as settled fact that the fifty-centimeter laminated carbonate sediments between the impact breccia and the KPB mass extinction was the backwash from the impact-generated tsunami. As before, any evidentiary inconsistencies with the impact theory were conveniently attributed to this mega-tsunami event.

Of course, my small team and my undergraduate students had proved the opposite on that blistering hot day in March 2000, with our own discovery of the primary impact spherule fallout at El Peñón, Mexico. The evidence there showed that the impact predated the mass extinction by about 200,000 years. But experience had shown me that we were unlikely to get our discovery published, and even if we did, we would be trashed into oblivion by the impactors. I knew the best way for us to convince open-minded people that the impact theory was wrong was to fight them on their own turf—the Chicxulub crater. So we temporarily put work on El Peñón aside and concentrated on the Chicxulub crater core. Would we find evidence of the true age of the impact crater within that core? I was confident we could. We just needed to get our hands on the samples.

It wasn't until May 2002 that I learned that Jan Smit had returned to Mexico and made off with the entire Yax-1 core sample for the KPB. I'd just come in from my garden in Princeton and decided to take a quick peek at my emails before dinner. I immediately lost my appetite. Considering the upcoming EGU-AGU grand conference in April 2003, and the importance of the Chicxulub impact core samples to our presentation, the news that Smit controlled the critical samples alarmed me. My team and I had been told that to secure a spot on the all-important conference program, abstracts were due by January 15, 2003. That wasn't far off. Smit promised to deliver samples to the other scientists no later than August 15, so each team could do research in time to present their results in Nice. Given his and his allies' mismanagement of vital scientific evidence in the past, I was more than a little skeptical.

The jackal, I thought to myself. If I hadn't already seen so much skullduggery in my years embroiled in this conflict, I wouldn't have believed it was possible. How could Smit, the same man who had been caught red-handed attempting to smuggle the core away in his duffel bag just weeks before, now be the custodian of the evidence? It was a terrible insult to the Mexican team. *He'd better not set foot in Mexico any time soon*, I thought, suppressing a grim laugh. I skipped down the hall to share the story with Majda, not yet realizing how this fateful development would alter the next months of our lives.

August came and went, and no Chicxulub samples arrived. The students returned to campus, providing a welcome distraction. I listened to them chatter in the halls as I gave the plants along my office windowsill a happy splash. Once again, my lab pulsed to the bright thrum of my grad students' music and the air filled with the spicy aromas of their homemade Indian food, reheated in the microwave in the corner.

Then September came. Still no samples arrived. At this point I was peppering Smit with frequent emails asking him what was going on. Smit answered my anxious requests for the samples with vague references to

130 THE LAST EXTINCTION

being "too busy," and promised that they would be delivered "next week." I tried to give him the benefit of the doubt. This was, after all, a busy time of year for everyone. I pushed away my growing anxiety.

Weeks passed. Now it was October. Outside the lab's windows, autumn sunlight burnished the leaves of the ancient elm a brilliant gold. In November, the leaves dropped. Still no samples. At this point, Wolfgang, Thierry, and I were checking in with each other on a near daily basis. Wolfgang was optimistic, predicting the samples would arrive any day. Thierry, ever the skeptic, declared the samples would never arrive. As the weeks passed by in November with no samples, it began to seem that meeting the looming January 15 deadline would be impossible.

We were in a race against time and knew we had to be organized and ready to roll as soon as the samples came in, if they ever did. I did my best to remain positive. On a call with Wolfgang and Thierry, I predicted that the samples would arrive just in time for study, and analysis of the critical core interval would support the pre-K–P age of the impact.

Wolfgang chuckled. "I'd love to see the faces of the impact guys when we present that evidence."

"Yes, so would I," I replied. "Although it'd be the last thing we ever see, because they'd kill us."

Thierry was more subdued. "I don't believe it's going to happen," he said. "The chances are very small that we will get the samples."

I raised my eyes skyward. In my mind I could see him rubbing his forehead and pulling on his thinning hair in characteristic Thierry fashion.

I took a deep breath, resolving myself for a more confrontational approach toward Smit. "I'll make sure we get the samples. The question is when." Then I paused for a moment, considering all possibilities. "And what if the results don't go our way?" I asked.

Thierry's response was immediate. "Then we see what additional analyses can be done for maximum information," he said.

Maximum information. I hung up the phone, still turning the phrase over in my mind. Maximum information had always been my objective. The spirit of our team was collaborative and cross-disciplinary by design. We were interested in where the facts enlisted scientists to weigh in, like geochemist Doris Stübenand Utz Kramar, climate scientist Zsolt Berner,

RACE TO THE DEADLINE 131

and sedimentologist Bill Ward, among others. Sure, Wolfgang, Thierry, and I had our theories, and we were passionate about our ideas. But we were most passionate about finding the truth, whatever that may be.

By contrast, the impact theorists never invited scientists outside their group to collaborate. To preserve their claims, no contrary evidence could be tolerated. If we stood any chance of beating them back, we had to amass a depth of data so convincing across disciplines that the truth could not be denied.

At that moment I devised a plan. As soon as the core samples came in, whenever that might be, I decided that we'd perform seven independent lines of investigations, across disciplines, to gain maximum insight into the age and depositional environment of the Chicxulub impact. When I shared this plan with Wolfgang and Thierry, they loved it, even though it was absurd to plan such an all-in approach when we didn't know we'd ever get the samples.

Our team already covered at least seven disciplines relevant to this core study, including fossil tracks and traces that reveal the seafloor environments; sediment structure and grain size to differentiate high-energy storms versus laminated sedimentation during quiet waters; chemistry of clay minerals precipitated from seawater during quiet times over thousands of years; foram fossils for control of the environmental conditions; the response to climate change; iridium concentration and other rare earth elements indicative of cosmic input. I contacted each of these scientists to prepare them for an intense period of research, should we get the sample in time.

And so, team assembled and poised at the ready, we waited. As the holidays approached, the students began to depart for the winter break, draining the halls of energy and cheer. When the first snowflakes of December glazed the skylight windows in the lab, I finally lost my cool and flew into a panic. I had again underestimated the depths to which Smit was willing to sink to protect his theory from scrutiny. I'd expected the petty delaying actions, but I never believed Smit would outright withhold the samples. Now that I saw his game for what it was, a deliberate and shameless sabotage, it was too late. How would we make the January 15 deadline now? The impact theorists would run away with the

132 THE LAST EXTINCTION

conference, declaring once and for all that the Chicxulub impact was KPB age and solely responsible for triggering the mass extinction. My team's absence would give the impact camp the glamorous and jubilant celebration they craved.

It was well past time to take control of the situation. I had one ace in my sleeve: the threat of calling NSF's Program Director Leonard Johnson, who supported the drilling project. I knew he would be furious to hear that Smit and the other PIs were going against the NSF rules that explicitly required them to provide samples to independent investigators that had been funded. Wanting to spare him the drama of this intramural squabble, I had yet to inform him of the situation. If my strategy didn't work, it would be my next call after Smit. I was convinced that even Smit would be worried about closing off the NSF funding spigot.

I placed a frantic call to Smit. *Breathe*, I coached myself as the phone rang. *Stay calm*, I told myself as he answered.

"Where are my samples?" I shouted. So much for staying calm.

"You will get them eventually, even if not in time," he replied. Did I detect satisfaction in his voice?

I took a steadying breath. "Jan," I said softly. "Don't you realize you and your entire drilling team are courting deep trouble?"

"How so?" he asked.

"Leonard Johnson, director of the US Continental Drilling Program that funded most of the Chicxulub drilling project, is furious."

He met my bluff with stony silence.

Sweetly, I pushed the knife in a little further. "Jan, surely you know that Johnson funded my proposal expressly to study those K–P cores. He knows you and the other five principal investigators of the drilling team are withholding the samples, and he is madder than hell. He won't be kind to any of you."

More silence. Finally, Smit said, "I will ship the samples express to you tomorrow."

December 22 was our lucky day. A small box wrapped in brown paper and plastered with stamps from the Netherlands arrived in my lab with Smit's Amsterdam University return address. I grabbed a penknife and opened the box, my heart racing. There they were, twenty-three samples

RACE TO THE DEADLINE 133

spanning the fifty-centimeter critical interval between the top of the tumbled impact breccia rocks and the K–P boundary clay, plus a few samples above. Each sample consisted of a small cube of white to gray hard limestone, thinly laminated and placed within a labeled plastic bag. A list of the samples was included. Nothing else.

My heart sank. The samples were very small, each about the size of a sugar cube. It was far from ideal, but it was something. By God, I thought, we'll do our damn best. I did a quick mental calculation. We had twenty-one days, including all holidays, before the abstract deadline on January 15. Could we give up the holidays? Majda and I had made our usual plan to head to our apartment in New York City, and he would be terribly disappointed. I thought of my collaborators; several had small children and surely could not abandon their long-anticipated celebrations. Could they be persuaded to drop everything and work round the clock? There was only one way to find out—call them, persuade, and set the stage.

That same day, I took my box of Chicxulub core samples and went looking for George Rose, the kind of unsung, superb craftsman who keeps labs running. I begged him to take on a monumental job on short notice—to cut and grind thin rock slices from each of the twenty-three samples. He knew all about the Dinosaur Wars and the importance of this evidence and he began preparations of the samples immediately, returning the unused rock samples the next day. I then split those samples and FedExed them to my collaborators Doris Stüben in Germany and Thierry in Switzerland. On Christmas Eve, I went to campus to pick up my thin sections from George Rose and with deep gratitude handed him a bag of dark Swiss chocolates, which he accepted with a sunny smile. He headed home to celebrate the holidays with his family, and I resumed the lonely climb to my lab on the third floor. Majda had given his blessing, if begrudgingly, to my need to spend our holidays in the lab.

The old building was deserted, and the heat turned down, making my large lab space painfully cold. The ancient hot water heating system never worked well in any case—my students and I often wore coats, hats, and mittens during the winter months. On this day, an icy wind blew

134 THE LAST EXTINCTION

through four drafty windows that were over a century old and a large skylight on the sloped roof. I turned on the space heater by my feet and wrapped myself in an old aqua-green wool coat left by one of my students. Outside, snow swirled and collected heavily on the branches of the elm. It would be a white Christmas.

I sat down to work in front of a long oak table repurposed from Princeton's original labs nearly 270 years ago. Then, I positioned the microscope and laid out my thinly cut core sections. Not for the first time, I wondered if I was on the verge of unlocking the great mystery of what caused the mass extinction. I hesitated and pushed the thin sections aside; it would be easier to first identify the forams washed out of the sediments.

I processed samples all Christmas Day and the next day in my cold lab, washing the fine fraction through a sieve with icy cold water and freezing my fingers stiff. When the water passing through the sieve was no longer muddy but clear, and the forams were clean, white, and visible as a tiny residue, the sample washing was complete. I rinsed the white residue through a filter paper lining a funnel, then placed the filter paper in the oven to dry.

Despite my freezing fingers, I didn't mind the manual aspect of this work. With enough repetition, the task becomes almost meditative. Sometimes I imagined the forams as living creatures, floating and scuttling across the sands of a prehistoric ocean. The repetitive action helped calm my thoughts and focus my attention. My mind tended to race ahead, this time dwelling overlong on fears that I wouldn't succeed. Then it would boomerang to thoughts of success with excitement and anticipation: *What if I actually pulled this off?*

Once the sample residues were dried in the oven, I took them home to study in my warm and comfortable home office, which was decorated with a myriad of mementos and art collected on my field trips and vacation travels. Majda was happy to have me around. His home office was just off the first-floor foyer, although he largely avoided it and instead stacked piles of research papers for different projects hip high, in rows along the walls. He preferred to commandeer the large dining table with a view of the two-story living room and wide windows to the garden, open land, and forest beyond.

After twenty-five years together, we had our routines down. We didn't talk between our downstairs-upstairs workplaces but occasionally checked in on each other. I rarely visited him because he was too absorbed in his work and didn't want to be disturbed. Majda visited me whenever he took a break. I was always happy to see him, but that December I had little time for visits.

I began my research at home with great anticipation. I pulled my chair up to my desk, turned on the microscope light, spread the first washed residue onto a black metal tray, and searched the tiny, washed residue for forams. Based on the other evidence my team and I had discovered that showed life after the impact event, I expected to find forams in these samples. Nothing. I tried the next slide. Same. Then the next. A sick, queasy feeling slid through my stomach. Sample after sample, the search was negative—no forams. My anticipation crashed and a dark mood enveloped me. I pushed away from my desk, cursing. I was so stupid. How could I have believed I'd ever find diagnostic forams in such tiny samples? I banged my fist on the table in a burst of anger.

"Gerta?" Majda called up the stairs. "What's all this?" He climbed the stairs and stood in the doorway, where I greeted him with a torrent of complaints.

"I've just wasted two days of miserable lab processing for nothing! Even worse, I've squandered the last bits of sediment samples that could have been used for other analyses." I cursed again.

"Calm down," he said. "This is no big deal. You still have the thin sections to analyze, and they may give you the data." I ignored him.

"Everyone's counting on me," I said. "I convinced them all to give up their holidays, and here I've badly misjudged the samples."

Majda listened to me rant, cupping his chin in the palm of his hand. He encouraged me to be optimistic, but I would hear none of it. I was done with hope. I needed space to release my frustration the old-fashioned way: dig a hole, chop down a tree, take a baseball bat to something, break some plates, or speed-run to exhaustion. These were all time-tested ways to

136 THE LAST EXTINCTION

release tension, clear the mind, and find your way out of a predicament. It works.

I blazed past Majda, and he trailed me down to the first floor, trying the whole way to dissuade me from going into the garden in the freezing weather. I ignored him and started pulling on additional layers: double sweatpants, sweater, two old coats, two pairs of gloves, a hat, heavy duty socks, and winter boots. In the garage I grabbed a pick and strode out into the forested part of our garden which had been invaded by dozens of red bud saplings. I swung the pick at the red bud roots. When the pick struck the frozen ground, tremors reverberated through my body. I quickly worked up a sweat. Majda watched me from the window of the summer greenhouse and shook his head, yelling, "You are crazy! Stop it!"

No, I mouthed back, my brain ringing from the sharp strikes of the pick.

He waved for me to come back inside the house.

I turned my back to him.

For the next two hours, I beat the frozen soil, digging out redbud saplings.

After the brutal release of tension in the frigid frozen garden, my mind was calm and clear—the freakout was over. I realized even if no age diagnostic forams were found in the washed samples or thin sections, this project still had many other lines of evidence that could prove the Chicxulub impact predated the mass extinction. And for now, I still had to analyze the thin sections that could hold the clues—even Majda had realized that. I went back into the warm house where Majda had cooked his special spaghetti with tomato sauce, red peppers, and Italian sausage with broccoli rabe. He looked at me apprehensively and then let out a deep sigh of relief when I laughed and gave him a hug. After dinner, we rewatched the classic low-budget 1974 horror film *It's Alive*, which we first saw shortly after its release and now laughed our way through it.

Back in the lab the next day, I felt strangely hopeful and ready to take on the day. I began with sample number twenty, taken just six centimeters above the impact breccia at the base of the fifty-centimeter critical core interval. I placed the slide on the microscope stage and turned on the light. To my astonishment, the sediment was well preserved, consisting

RACE TO THE DEADLINE 137

of millimeter thin light and dark layers. These clearly demarcated layers meant normal seasonal sediment deposition had returned just a short time after the Chicxulub impact. I scanned the glass slide for disturbed deposition, which might signal storms or current erosion but saw none. The sediments were recrystallized but not as totally as in the underlying samples. I came across a less recrystallized pocket, inhaled sharply with anticipation, and squinted into the scope.

They were there. The forams. I gasped. It was a miracle! I shot up from my chair, my heart pounding, mind racing. Almost delirious, I pranced around my lab benches, alternately laughing and clapping my hands, in semi-disbelief. I took another look to convince myself that what I saw was real. It was. Within the less recrystallized pockets, I identified thirteen foram species from an assemblage that had lived long before the KPB mass extinction. Most important among these was *Plummerita hantkeninoides*, which evolved 220,000 years ago and went extinct at the KPB sixty-six million years ago.

Evidence taken directly from the Chicxulub crater revealed that the impact occurred a very long time before the KPB mass extinction. It was the proof we'd been looking for. It confirmed the discovery at El Peñón in 2000, and Loma Cerca in 1999 where the primary spherule fallout revealed that the Chicxulub impact preceded the fifth mass extinction by about 200,000 years. We had laid this stunning discovery aside to focus on the Chicxulub crater core to match what we'd already found. When multiple lines of evidence start telling the same story it becomes harder and harder to deny it.

At home that evening I shared my findings with Majda, who was happy. He then asked numerous questions to test my findings. After about an hour he was satisfied, sat back in his chair, and said, "G, you've got it!" A deep sigh of relief escaped my chest. I passed my most critical test.

The next day, January 7, I called my collaborators in Germany and Switzerland to reveal the exciting news: The age of the Chicxulub impact crater predated the KPB and confirmed the age of the primary impact spherule fallout at El Peñón and Loma Cerca. What clinched it was the presence of the foram *P. hantkeninoides* in the critical fifty-centimeter

138 THE LAST EXTINCTION

interval between the impact breccia and the KPB. My collaborators' excitement was palpable. We laughed with pure joy, nearly uncontrollably, releasing some of the accumulated tension, not only from the past week, but the past ten years. I was wiping tears from my eyes from laughing so hard.

Over the next few days, my colleagues reported their own findings, each one adding more supporting evidence of long-term environmental changes that supported my age determination. Wolfgang's investigation of trace fossils revealed feeding traces and burrows, which couldn't exist if sediments represented tsunami deposition over hours to days as claimed.

A thrilling call with Doris Stüben and Zsolt Berner revealed that nutrient cycling in the ocean, which is nature's recycling system, remained high during the fifty-centimeter core. And then abruptly stopped due to sediment erosion—a hiatus. Above the hiatus, dark gray clay marks the KPB transition and early Paleocene. What had just been revealed was new information about the sediments below and above the KPB, which included: the incomplete sedimentary record of the fifty-centimeter core below the Chicxulub impact breccia, a major hiatus at the top of the core that misses much of the late Cretaceous sediments, and the KPB transition that is also partly eroded.

Thierry added more good news. He analyzed the large grain sizes in the critical fifty-centimeter core interval. Smit contended that these were bits of sand introduced by the impact tsunami, but Thierry showed that they had crystallized over millions of years. Above this interval, five thin green clay layers of glauconite revealed slow deposition over tens of thousands of years in quiet waters. Smit believed the green layers were clay-altered impact spherules. There was no impact glass. One by one, the impactors' tsunami interpretation was wrong.

A new wave of excitement coursed through me. We had it all. The proposed "impact-tsunami" was normal sedimentation, and settling from the water column, and precipitation on a quiet seafloor. The Chicxulub impact occurred about 200,000 years prior to the KPB mass extinction, a fact confirmed by our earlier discovery of the primary impact spherule layer at El Peñón that could also be dated 200,000 years prior to the KPB mass extinction in northeastern Mexico.

RACE TO THE DEADLINE 139

There was one more independent age confirmation we knew could strengthen our case: the geomagnetic reversals of Earth's magnetic field. A geomagnetic reversal occurs when the magnetic north and magnetic south switch places, a phenomenon that occurs during times of major volcanism and plate tectonic activity. The KPB mass extinction occurred near the middle of Earth's twenty-ninth geomagnetic reversal, which lasted about 750,000 years. Finding evidence of this polarity reversal in the critical fifty-centimeter core would confirm what we already knew from other localities: The foram zone CF1 is in the twenty-ninth reversed polarity below the KPB.

We knew just the right people to ask for this collaboration. Jaime Urrutia-Fucugauchi, one of the PIs of the Chicxulub crater drilling team, and his colleague Mario Rebolledo-Vieyra, both from UNAM. They were experts on geomagnetic reversal dating and had already measured the Yax-1 core. We had friendly communications in the past and this collaboration could be mutually beneficial.

My email to Jaime and Mario was friendly, though carefully worded. I needed to feel out whether they were willing to risk collaborating with my team. I informed them I was cautiously optimistic that their work might also support a pre-KPB age for the critical fifty-centimeter interval and a major missing interval (known as a "hiatus") above. I took a deep breath, hit send, and waited. Their reply was almost instantaneous. They were happy to collaborate because the data they had found fit my foram data precisely and they needed it as independent confirmation to publish their own paper.

This was a major boon to our study because a principal member of the Chicxulub drilling team had data unequivocally supporting our results. The impact theorists would view their cooperation with my team as a stunning betrayal.

In the final days leading up to the abstract deadline, more results from my collaborators trickled in, each providing data that confirmed the same story, albeit from the perspective of different disciplines. I received magnetic reversal data from Jaime and Mario; carbon isotope data and iridium concentrations from Doris' team; evidence of seasonal sediment deposition and seafloor burrowers from Wolfgang; and five layers of

140 THE LAST EXTINCTION

green clay, each formed over tens of thousands of years from Thierry. All these different studies confirmed what the foram data had revealed: The Chicxulub impact predated the mass extinction by approximately 200,000 years.

From where I stood, the evidence was incontrovertible: The impact did not trigger the mass extinction. It was a complete affirmation of our previous work's conclusions, which had been so brutally denied and suppressed by the impactors. I hungered for the chance to present our findings in Nice.

On January 15, I sat down to write the abstract for submission to the EGU-AGU conference in Nice, France, to present our data at the unveiling of the *Chicxulub Impact Drilling Results Session* organized by the drilling team. I wrote carefully within the prescribed limit of 350 words, focusing on core sediment description and its implications. I kept it as dry as possible, revealing no new data that might lead the organizers of the session to reject us. I concluded that the evidence pointed to an extended time interval of deposition that suggested the Chicxulub impact may have predated the mass extinction.

Wolfgang and Thierry also submitted abstracts on their studies.

Days later, we learned the news. All three abstracts had been accepted.

The stage was set for the showdown in Nice.

11

SHOWDOWN *in* NICE

SHORTLY AFTER 6 A.M. ON APRIL 6, 2003, I MADE MY WAY
toward the Nice Conference Center. Early sunlight bathed rows of houses
in a warm glow and chased away the cool night air. I felt calm and prepared. It was the most important day of my entire science career, and I
was ready.

I had only crossed a few blocks when muffled shouting and crowd
noise pierced the morning quiet. As I turned onto the main boulevard,
the cries grew more distinct: "Down with the USA! Down with the
USA!" I recognized the sounds of a large demonstration and surmised it
must be a protest against President Bush's recent invasion of Iraq, a move
which set off worldwide condemnation. Turning the corner, I gasped and
pulled up short. Hundreds of people crowded in front of the conference
center. Thousands more lined half the street and stretched block after
block well beyond sight.

Thierry and Wolfgang appeared by my side. We had arranged to meet
at the conference early so that we would have plenty of time to prepare
before my 8:20 a.m. talk. Now, assessing the crowd, I realized that we
might have trouble making the talk at all. We hadn't seen each other
in nearly a year, had rarely talked since our harrowing but successful
three-week investigation of the Chicxulub core over Christmas and the
New Year. I was eager to catch up with them, but the swelling crowd filled

142 THE LAST EXTINCTION

the air with tension, and after a quick greeting we wordlessly made our way toward the packed entrance.

In the wake of the September 11 terrorist attacks in 2001, there was heightened security everywhere. We speculated that measures might be particularly severe that morning due to the anti-US demonstration. This, after all, was the first-ever joint meeting of the European Geosciences Union and American Geophysical Union (EGU-AGU). The protesters were likely targeting the strong American presence at the conference.

Two security booths had been set up outside the doors to the conference halls. We watched scientists approach the booths one by one to have their bodies, bags, and computers checked before entering. The process was painfully slow. I grabbed Thierry and Wolfgang by the arms and together we speed-walked down the street to find the end of the line. Our jaws dropped as we passed beneath immense banners promoting the Chicxulub impact crater drilling results as the grand event of the conference. This was going to be an even bigger deal than we expected. We walked on in silence, intimidated and breathless, and eventually made it to the end of the line some eight blocks away.

The Chicxulub Symposium was set to begin at 8 a.m. Jan Smit would give the first twenty-minute presentation, and mine would follow him for the next twenty minutes. I looked at my watch. We had one hour before my talk.

The line to get through security moved slowly, imperceptibly, as the minutes ticked away. When I realized we'd moved no more than one meter in thirty minutes, I almost screamed. There was no way we'd make it into the conference center for the presentation of our results. After our relentless drive to conduct research and submit the abstract, the sacrifice of the holidays, and the strains we endured in our family relationships, we were about to lose our once-in-a-lifetime opportunity to present our work to the largest international scientific audience.

It was now 7:45 a.m., just fifteen minutes from the official start of the Chicxulub Symposium. My heart racing, I turned to Thierry and Wolfgang.

"I've got to do something to get in. This is our chance to make a difference. Wish me luck."

SHOWDOWN IN NICE 143

I sprinted down the blocks to the entrance, where hundreds of people clamored to get the attention of the security guards, pleading their cases to be let in. In the commotion of anger, shouting, and shoving, I managed to squeeze through the crowd and made a beeline for the guard at the defunct booth.

Summoning my most angelic smile, I asked, "Can you please let me through? My lecture starts in fifteen minutes."

He sized me up, no doubt assessing my threat-level. I held my breath and imagined my appearance through his eyes: a petite, neatly dressed middle-aged woman with a mild Swiss accent. Judging me sufficiently harmless, he waved me through. With my heart hammering in my ears, I thanked him and rushed to the grand auditorium. The room had a seating capacity for about 1,000 people, but I opened the door to a sea of empty blue seats. Down by the podium I saw a cluster of perhaps twenty-five people huddled together—the Chicxulub impact drilling team.

There was only one friendly face: Professor Jason Morgan, my Princeton colleague, who was sitting alone halfway up the rows of seats. He nodded and smiled encouragement. About a dozen people sat dispersed through the auditorium, press tags on their lapels and notebooks in hand—the journalists. Down by the podium, the Chicxulub team members talked noisily. Someone spotted me and a rumble rose from the group. A wave of hostility hit me as I walked down the aisle, their stares burning holes into my body. I chose the first row on the right side as far away from them as possible.

Let them stare. They didn't bother me, but the nearly empty grand auditorium did. I'd hoped I'd be addressing a large group of persuadable scientists, but I realized I would be giving my talk to two dozen impactors, a dozen journalists, one colleague, and a few more scientists who might still trickle in. The impactors were masters at controlling opposition science and condemning them to obscurity. I feared without the broad audience and large media attendance it would be too easy for them to deny, dismiss, and vilify me to those who couldn't make it to my presentation. If we succeeded in getting the story out, we could win. If not, our results would never gain recognition and would be lost to history.

144 THE LAST EXTINCTION

Of course I knew the Chicxulub symposium would be a showdown—only now I realized I was in for a far nastier experience than I anticipated. With the audience dominated by the Chicxulub drilling team, this could easily turn into Snowbird III all over again. I knew my talk would enrage them, and without an objective audience in attendance, I could easily become their piñata for the day. So, what now? Come what may—I would present our evidence. Let them scream. We had the data.

At 8 a.m., Smit confidently strode up to the podium and began summarizing the drilling team's conclusions. The asteroid hit Yucatan at precisely the K–P boundary and caused the mass extinction. The signature of this impact was the jumble of partly melted rocks known as impact breccia. The fifty-centimeter interval of limestone in the Yax-1 core that separated the impact breccia from the mass extinction was caused by impact-generated tsunami waves that backwashed and infilled the crater . . .

Listening to Smit's tale, I was reminded of *Groundhog Day*, the 1993 comedy about a weatherman who finds himself living the same day over and over again. What is initially a novel experience turns into a nightmare, as the weatherman is forced to experience the same events, day after day after day, until he can break the spell. My Groundhog Day began in 1992 when I first heard Smit apply Jody Bourgeois's 1988 impact-tsunami scenario—a theory initially proposed to explain KPB mass extinction sections found near the Brazos River in Texas—to explain the meters of sandstones and mudstones that separated impact glass spherules from the KPB mass extinction in northeastern Mexico. Now at Nice in 2003, Smit trotted out the same tired impact-tsunami scenario to explain the fifty centimeters of limestone that separated the impact breccia from the K–P mass extinction layer. And, once again, he failed to do the necessary analyses to test that his visual interpretations were right. His ideas were no more plausible than in 1992. Hopefully today, I thought, I could end this tiresome loop.

Facts don't lie. But lies can be invented easily and wedded to facts. Smit was a master at weaving believable scenarios that blended facts and lies to fit the impact theory. He based his story on simple, well-known principles: that tsunamis can erode, transport, and redeposit sediments;

that earthquakes destabilize slopes and cause gravity flows and slumps; that waves create backwash from shores and infill craters. It all sounds very plausible until you wonder why so many key pieces of evidence conveniently show signs of unusual geological activity at just the right moments.

When Smit finished, there was wild clapping, and a few shouts and whistles erupted from the exuberant Chicxulub drilling team. One almost had to admire his brio. I wondered, had Smit really looked at the same core segment that we analyzed? There was so much evidence he missed or ignored, so much that was misread and misrepresented. I could scarcely keep the grin off my face.

Now it was my turn. As I walked to the podium my first slide flashed on the screen: "Chicxulub—the Non-Smoking Gun: Impact Crater Predates K–P Boundary." Audible groans arose from the front rows before the podium.

"Not that again," somebody hissed.

Another yelled, "It's always the same."

"Nonsense! All wrong."

A measured voice rose above the others, asking for quiet. "Let Keller give her talk and ask questions at the end." It was Buck Sharpton, one of the drilling team members.

I looked at the sparse audience, now growing as another dozen scientists trickled in, and began. "You've heard Jan Smit's interpretation. My team analyzed exactly the same core interval, exactly the same samples, and you will now hear exactly the opposite conclusions.

"The Chicxulub impact predates the KPB mass extinction," I continued. "There is no tsunami deposition. We base our conclusions on seven independent lines of evidence and all concern the same fifty centimeters between the top of the impact breccia and the KPB boundary."

Upon hearing this I noticed a few faces among the riled-up impact group register alarm.

"Is it normal sedimentation or tsunami backwash?" I asked rhetorically. "All seven independent characteristics of the fifty-centimeter-long Yax-1 Chicxulub crater core reveal normal sedimentation." Now I had everyone's attention.

146 THE LAST EXTINCTION

Methodically, I walked the audience through our evidence. I explained how Yax-1 consisted mainly of laminated limestone, with thin light and darker tan layers corresponding to dry and rainy seasons, indicating normal long-term sedimentation. Then I described the carbon isotope composition of the limestone, which again showed normal high levels during the end-Cretaceous and very low levels after the mass extinction. Entirely absent were any chaotic high and low signals from mixed rocks of different ages, and types as would be found in jumbled tsunami backwash deposits.

So far, the crowd remained attentive. I wondered for a split second whether I might make it through my talk unscathed. Not a chance.

"During any storm or tsunami event, the seawater is turbulent with suspended sand. When the waves subside, first the heavier, larger sand grains settle to the seafloor and subsequently the smaller ones. This is Smit's evidence of the tsunami event in the fifty-centimeter Yax-1 core." I let this sink in for a moment, then continued: "But in fact, that's not the case." Titters emerged from the gallery. I continued: "The grains he referred to are magnesium carbonate crystals, known as dolomite, that formed by chemical alteration in limestone. *These crystals were never suspended in the water column. There is no evidence of a tsunami event.*"

At these words, Smit cried out, "Impossible! These are sand grains from the tsunami."

I politely deferred questions to the end of my talk, knowing that if I stopped to address any criticism now, they would never let me finish.

Next, I demonstrated that the chemistry of the five green clay layers in the upper thirty-seven centimeters of the fifty-centimeter Yax-1 core is not impact glass, as Smit claimed. Based on the chemical composition, it is green glauconite clay, an iron potassium phyllosilicate mineral. Each of these layers formed slowly on a quiet seafloor over tens of thousands of years. This again ruled out a tsunami.

Agitated whispers arose from the crowd as the impact team exchanged worried glances.

Calmly, I pointed out the next lethal evidence for the tsunami deposition: the burrows and feeding traces of the abundant marine life that had colonized the seafloor in the fifty-centimeter core interval. This ruled out tsunami backwash or any other rapid deposition that would bury seafloor life.

While pictures of the burrows flashed on screen, Smit continued to lose his cool. "This is nonsense. There is not a single burrow! She doesn't know what she is talking about!"

Buck stood up and demanded, "Sit down, Jan, and let Gerta speak."

Feeling momentum, I cruised into the line of evidence I knew best: my planktonic foram analysis. With the air electric with tension, I announced my discovery of *Plummerita hantkeninoides*, a species that indicated the fifty-centimeter core interval was deposited in normal quiet waters during the last 200,000 years before the mass extinction. It was the same age as the primary impact spherule fallout we had discovered in northeastern Mexico.

Smit had denied the presence of any forams in the sediments, claiming those were dolomite crystals. I pointed out that he had mistaken dolomite crystals for sand grains in a tsunami and now claimed forams were dolomite crystals. No foram specialist with two functioning eyes could mistake dolomite crystals for sand grains and even fewer forams. These were bad, easily demonstrable mistakes. But I wasn't done: We had saved the most dramatic evidence for last. I pressed on, savoring what was about to come next.

I then introduced the magnetic reversal research done by the drilling team, co-PI Jaime Urrutia-Fucugauchi and his collaborator Mario Rebolledo-Vieyra, who dated the age of the critical fifty-centimeter core as the lower part of magnetic reversal C29r, which is also foram zone CF1 below the KPB mass extinction. As I described their results, my summary graph of all the evidence that illustrated C29r reversal agreed with all our age-related data in this core interval that flashed on the screen. This was the worst blow for the impactors. Not only had Jaime and Mario agreed to collaborate with us, but their data supported our pre-KPB age of the Chicxulub impact. It was now impossible for them to simply dismiss our data as biased. Members of the impact camp had shown the impact did not cause the mass extinction.

Smit and his allies erupted in fury. From the stage I saw them turn toward Jaime and Mario, pointing their fingers as they hurled accusations

148 THE LAST EXTINCTION

of disloyalty. I felt sorry for them. They had seemed proud of their collaboration with my team, and as I presented my analysis, I could see them nodding and smiling. Now they were shrinking into their seats as if seeking protection from the mob. I'm not sure they were aware of how vindictive their former allies could be when the impact theory was questioned. Jaime occasionally tried to answer but could not be heard above the din. For a moment, Jaime and Mario, rather than I, had become the target of attacks for nothing more than contributing their supporting science results to our study.

In the continuing cacophony, Sharpton's voice rose: "Calm down, let Gerta finish her lecture."

Fighting the din, I concluded. The Chicxulub crater drilling Yax-1 revealed the true age of this impact was not sixty-six million years ago but rather at 66.200 million years ago, which is 200,000 years prior to the KPB mass extinction. This age is supported by the recent discoveries of the primary impact spherule fallout at two sites in northeastern Mexico just before peak climate warming preceded Deccan volcanism in India. Despite the large size of the Chicxulub impact crater (roughly 175 kilometers in diameter), it didn't cause any significant species extinctions and left no significant long-term environmental changes, whether in the fifty-centimeter Yax-1 core or in rock outcrops surrounding the crater in a radius of approximately 2,000 kilometers. The real disaster struck sixty-six million years ago and caused the fifth mass extinction, which can now be attributed to Deccan volcanism. The impact theory requires modification.

I thanked the audience for their attention. Pandemonium broke out as the drilling team turned from Mario and Jaime and again unleashed their venom on me. As I walked off the stage, I wondered, had I succeeded in raising awareness among at least some scientists that the impact theory and KPB mass extinction was no longer valid?

I collapsed, exhausted, into my front-row chair far away from the impact crowd. A burly man dressed in baggy shorts and a colorful Hawaiian shirt rushed up, flopped down in the seat next to me, and handed me his business card. "I'm Rex Dalton," the man said, "an investigative reporter for *Nature*. You've got the story. Don't go away."

Who was this guy? I wondered. *Nature,* just like *Science,* exclusively supported the impact theory. And now a journalist who didn't understand science would report the impact theory as proven and sandbag me as the woman and her small team who challenged consensus science to no avail.

I sat quietly through the next talks, all of them by the Chicxulub drilling team on topics of impact crater physics and the types of rocks in the impact breccia. Each speaker began by conspicuously proclaiming the crater was proven to be precisely KPB age, as Smit had claimed. *This is damage control,* I thought, and it might just work since the slowly increasing audience was unaware of the early morning shootout.

I learned during a coffee break that Buck Sharpton had announced an unscheduled noon meeting to deal with the new developments and fallout of my talk. I was not invited. I also noticed that Thierry and Wolfgang still hadn't appeared. I missed them. It was a lot more fun to talk to friends than to be glared at by enemies. Rex reappeared next to me and said he would go to the noon meeting and see what they were going to do. He believed that as a reporter he had access. I shook my head. *Good luck,* I thought. *Merely sitting next to me turned this reporter toxic for impactors.*

Rex's excitement and confidence intrigued me. He had the appearance of a long-haul truck driver who just stumbled into the Nice Conference Center for some lively amusement. I later learned that was exactly his earlier profession before he became a investigative science journalist. I soon learned he was not only that, but a larger-than-life character who possessed a very quick mind and a vast understanding of science.

At noontime, Rex was barred from the Chicxulub impact meeting.

I was still sitting in the conference hall when Rex returned. Wolfgang and Thierry finally made it through the security check and showed up. Flushed with laughter, they sped down the aisle to meet me, and I wondered what had them so animated.

"You did it," Wolfgang said breathlessly. "Our evidence, your bombshell lecture, everyone's talking about it. Rumors of your talk and Smit's bad reaction are everywhere. We first heard about it in line outside the conference center. Everyone wants to know more."

150 THE LAST EXTINCTION

Thierry nodded, his usual sedate expression replaced by an enormous grin. I laughed with relief and joy. Soon we were all laughing, releasing the tension we'd been carrying for days.

Rex regarded us with amusement and invited us to have lunch at the press center. For the rest of the conference, he glued himself to our sides to keep away other reporters waiting for an opportunity to interview us.

The afternoon session began with the announcement of a gag order by Dieter Stöffler, the German co-PI of the drilling team: "All publications about the Chicxulub drilling results are forbidden for at least one year and until after the *Meteoritics and Planetary Science* volume on the Chicxulub drilling results are published." This highly unorthodox order was the decision of the unscheduled noon meeting called to contain our research results. The gag order also carried no force because it went against NSF rules which permit the publication of results nine months after the drilling ends—a deadline that had passed six months ago. To be sure, I asked Buck what he thought of this gag rule. "It's invalid," he said, "they have no right to prevent you from publishing your results." During the coffee break I asked Jaime and Mario whether this attack on them changed their minds about being co-authors on our paper. Jaime said, "Absolutely not."

Although only a few scientists could attend the morning session of the Chicxulub symposium because of the security fiasco, our core results became the talk of the meeting. Everywhere we went, scientists congratulated us for our work. Impactors regarded us with open hostility, but among this crowd they were the minority. Phone calls from TV news channels, talk shows, and reporters asking for immediate interviews rained on us. We never found ourselves alone. For Thierry, Wolfgang, and myself, the remainder of the meeting was a whirlwind. All three of us were occupied separately, showing and explaining our data to groups from different disciplines. Many were perplexed by the attempt by other scientists to suppress our findings. Ironically, the drilling team's decisive dismissal of all our multidisciplinary evidence piqued the curiosity of open-minded scientists and worked to our advantage to discuss and demonstrate the evidence.

SHOWDOWN IN NICE 151

We weren't the only ones talking. While we explained our multiple lines of evidence to fellow scientists in the packed conference hall, Smit was working hard in his corner to discredit the foram evidence. He asked well-known foram experts to say that the forams didn't exist, but everyone agreed they were there. Strangely, Smit didn't contest the seven lines of evidence we demonstrated in our lecture that were the talk in the conference hall. It was yet another Groundhog Day that Smit could never let go, playing the same game over and over again—until the spell broke, and the game ended.

Smit could not take defeat, nor could the hardcore impactors who refused to accept the real evidence that the Chicxulub impact crater and the impact spherules at El Peñón and Loma Cerca in northeastern Mexico predated the KPB mass extinction by 200,000 years. But many other impact scientists dropped out of the Chicxulub game, including some of the Chicxulub drilling PIs. The hardcore evidence we had presented was irrefutable.

Smit and Stöffler, however, regrouped and doubled down on their theory. Within a matter of months, the impactors claimed the Chicxulub crater drilling was done in error and must be redrilled. I had seen this story before, too. If the evidence didn't fit the theory, then the evidence must be wrong. With this claim, the impactors reaffirmed their faith in the old belief and the Chicxulub story could continue unchallenged. Soon, the impactors stopped mentioning the impact crater core. It was as if it never existed.

Rex Dalton invited us to dinner that night, asking us to recount the full history of the Dinosaur Wars. He was a very quick study, posed incisive questions, and took copious notes. In the subsequent months, Rex verified our stories. His investigative report "Hot Tempers, Hard Core" appeared in *Nature* on September 4, 2003.

During the last day of the Nice meeting, Rex persuaded Thierry, Wolfgang, and me to submit our paper on the Chicxulub core to *Nature* because he believed the editor, John VanDecar, would be more open to research that questioned the impact theory than he had in the past. I doubted his advice but agreed to try. We submitted the paper in May. By

152 THE LAST EXTINCTION

the end of August, we received the editor's letter of rejection, along with three peer reviews. We weren't surprised by the rejection but were puzzled by the two reasons cited for the rejection: one, that we hadn't made a sufficient case that the sediments weren't tsunami deposits; and two, that we hadn't reconciled our conclusions with other regional evidence that dated Chicxulub at KPB age. What was strange was that neither of these reasons was given in any of the three reviews.

In fact, the first review was positive. The second, while more skeptical, acknowledged our points and the validity of our research. The third review was by Alan Hildebrand, the impactor who in 1990 had misinformed me that the PEMEX cores had been destroyed in a warehouse fire. His review attacked me personally and was tellingly devoid of scientific content. In eleven single-spaced pages, he lambasted me as a "hysterical female" with an "emotional" opposition to the impact theory. He also bemoaned having too few references of his own publications. The six different lines of evidence our team had gathered were dismissed as "emotional grudge and not legitimate research." Any reasonable editor would automatically disqualify such a review. But VanDecar demanded detailed replies to Alan's "grudge." I refused to answer Alan's eleven pages of paranoid ramblings. We revised the paper addressing all serious science comments and resubmitted. VanDecar rejected it again, this time based on one single reviewer's new comment:

If Keller, et. al. "are right, the weight of the implications is too unfathomable. I just can't believe that so many could have been so wrong for so long."

For *Nature*, the sheer strength of the impact camp and the likelihood of incurring their wrath was too strong. Their loss was PNAS's gain.

We immediately submitted our study to PNAS, the most prestigious science journal of the National Academy of Sciences. We received professional, highly favorable reviews and the request to rush publication, including the one-page color photo of an impact crater.

Our article appeared on March 16, 2004, with a front cover depicting an artist's impression of the Chicxulub impact and the headline "Chicxulub Crater Predates K–P Mass Extinction." Our study was a tremendous success. The following year (May 24, 2005), the PNAS editor notified me: "The paper you published last year has been met with great enthusiasm. The PNAS website receives almost two million hits every week and in 2004 over ten million article PDFs were downloaded. Your article was among the top one hundred articles accessed."

Our last hurdle was publishing in *Meteoritics & Planetary Science* (*MAPS*). The Chicxulub drilling PIs ordered all research participants to submit their articles for publication in this journal, and all six of them acted as co-editors. On July 17, 2003, I dutifully submitted our article: "More Evidence That the Chicxulub Impact Predates the K–P Mass Extinction." My submission was not acknowledged, and I received no response to my repeated inquiries.

Ten months later, on May 18, 2004, *MAPS*'s editor informed me that I was well past the deadline for submitting my revised paper and it was now considered effectively withdrawn. Apparently, they had received it, and even had it reviewed, but no one had shared the feedback with us. I informed the editor I had never received reviews and had no intentions to withdraw our paper. I then contacted Jaime Urrutia, who sent me the two very positive reviews, one of which was from the Chicxulub drilling co-PI, Dick Buffler. I forwarded these to the journal's editor who accepted the publication. Then he sent me a personal note:

"I must admit the editing of the volume on Chicxulub has been a strange adventure, with several guest editors with many complex confrontational and contradictory opinions. I will not be using this particular model for editing a volume ever again."

Though I was sympathetic toward this bystander who had been caught in the crossfire of this decades-long conflict, I was also heartened by his words. Apparently, our research of the Chicxulub crater core had changed some minds. There was now internal division in the impact group. The open-minded scientists were questioning the true believers and their ever-wilder impact scenarios devoid of facts.

Within a year of the EGU-AGU conference, the Chicxulub drilling team was singing a different tune. Most had abandoned their claim that the core evidence indisputably showed that the impact was KPB age. They now maintained that the KPB transition in the Yax-1 core was incomplete and no age determination could be made. And so, the great impact crater

154 THE LAST EXTINCTION

drilling project came to an end, and not the one that the impact theorists had hoped for. Instead of settling the KPB age controversy once and for all, the result was a major crisis in the impact camp.

In the fall of 2004, BBC Horizon reported our story in the documentary, *What Really Killed the Dinosaurs?* This was the first (and only) educational program that portrayed both sides of the impact controversy and gave serious attention to another catastrophe scenario: Deccan volcanism. My collaborators, Wolfgang and Norman MacLeod, had great roles in it. The show positioned me and Smit as protagonist and antagonist, two old dinosaurs duking it out to win the debate. I would be lying if I said I didn't enjoy it. But in truth, I no longer dwelled on Smit. He was in my rearview mirror. I had done all I could to show that the impact theory was wrong about what caused the fifth mass extinction. But what theory was right? This question had already been a focus of my work for years. It was time to give that my full attention to the real story of what caused the KPB mass extinction—was it Deccan volcanism?

12

TIME *to Be* BOLD

AFTER THE NICE CONFERENCE, I TOOK SOME TIME TO REFLECT on life and my career. It was clear the impact theory wasn't going to just disappear, even though every scientific foundation on which it stood had been severely undermined by the research my colleagues and I had amassed and presented. I was wise enough at this point to know that some people will cling to their beliefs no matter how untenable they are, and this is especially true when one's entire career has been built around an untenable belief. I was resigned to the fact that it was going to take time for the misconceptions that had been spread to the public for decades to be corrected. We had done everything in our power to show the scientific weakness of the impact theory. There was nothing more to be done to disprove it. So why did I still feel restless?

I had built a happy home with Majda and was well established in my science career. But I also yearned for the freedom and adventure I'd experienced during those four years of world travels when I was young and daring, when I had to be quick-thinking and fearless. After Nice, I found myself reflecting often on these early adventures while shopping in the grocery store, digging in the garden, or hunching over the microscope. My feet remained planted in a realm of comfort and security, but my mind was a million miles away, dodging bullets in Libya or trekking through jungle war zones in Southeast Asia. Nice was my team's first

156 THE LAST EXTINCTION

major public victory after fifteen years of being hounded by the impact theorists.

When I chose a career as scientist, I had, for the most part, traded this impulsive lifestyle for something more conventional. It was a life based on steady incremental progress built upon the careful collection of data. I'd employed this conservative science approach successfully for nearly two decades establishing a solid base for an environmental explanation of what led to the mass extinction. I'd succeeded by carefully treading a middle path, never asking for too much or proposing projects that were too risky. This was the kind of low-risk science the NSF rewarded.

Looking back on the twenty-one days of frenzied research in December and January, I realized the fast pace and high stakes had forced me to work with an urgency and ingenuity that had been stifled during years of methodical research. In the frantic lead-up to Nice, I'd reverted to something of my former self. I had tasted again that rush of adrenaline that comes from trusting one's gut and making split-second decisions. And now I craved more.

This was a pivotal time in my career. My team had revealed that mass extinction could *not* be linked to the Chicxulub impact, which multiple lines of evidence revealed preceded the KPB by about 200,000 years. But proving that theory wrong still didn't answer the question: What really happened during the fifth extinction? Perhaps, I reasoned, it was time to put this restless energy toward the question I had been wanting to answer since I was a graduate student. It was time to move on to investigate the alternative mass extinction hypothesis: Deccan volcanism.

At this point I was fifty-eight years old, and I wasn't getting any younger. If I wanted to solve the mystery of the fifth mass extinction in my lifetime, I would need to rediscover the restless spirit that had served me so well in my youth. It was a liberating revelation. I decided to mark the occasion by taking my two most adventurous sisters, Rosie and Helen, on a trip to spend time with Guatemala's indigenous peoples, the Maya, in the Western Highlands. It was their first trip beyond the confines of western Europe. We spent the days visiting villages around Lake Atitlán and admired the beauty of the Maya people's artful colorful clothing. We reveled in their large and thriving markets at Sololá and Chichi, where they sold everything from farm goods to traditional embroidery by different tribes, wood carved animal masks, and voodoo puppets. At nights, from the safety of our hotel garden in Antigua, we were mesmerized by the fireworks of nearby volcanic eruptions glowing red and orange. I tried to explain to my sisters how volcanic eruptions like these, multiplied by a million, could cause mass extinctions across the entire planet. It was one of our most exotic trips and we loved every minute of it.

When I returned home, I immediately planned my next steps to investigate what role Deccan volcanism might have played in the mass extinction. This required me to do three things: learn more about the science of global climate warming and the killing effects of major Deccan eruptions in India, locate evidence of the mass extinction in the 3,400 meters high basalt flows of the Deccan Traps, and date these basalt flows to establish

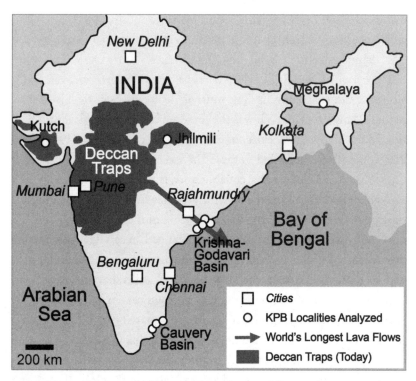

India's Deccan Traps region today. The original Deccan Traps covered three times this area across India. The longest lava flows originated in the Deccan mountains near Pune in the Western Ghats and flowed through lava tunnels more than 1,000 kilometers to the east to Rajahmundry and into the Bay of Bengal.

158 THE LAST EXTINCTION

a link to the global mass extinction. This was an ambitious and daunting list of future investigations. So far scientists had only made cursory studies into Deccan's contribution to global climate warming, and nobody had definitively linked Deccan volcanism to climate change.

Part of the reason for this lack of research was the fact that geoscience journals, so long under the influence of the impact theory, had shown little interest in Deccan volcanism. In 1998, for instance, one of my graduate students, Lianquan Li, produced a very high-resolution climate study of the South Atlantic Ocean that revealed abrupt deep-sea warming at the end of the Cretaceous. We suggested it was rapid climate warming caused by Deccan volcanism. Reviewers praised the research but demanded we cut any mention of Deccan volcanism. At the time it was taboo in the impact world.

There were a few scientists doing work in this area. In 1992, for instance, while I was studying foram fossils found in El Kef, Tunisia, I recall receiving an exciting new study from India. The main authors were B. C. Jaiprakash and D. S. N. Raju, from India's Oil and Gas Corporation (ONGC), which had drilled deep cores into the Krishna-Godavari Basin along the east coast of India. The Journal Geological Society of India had requested my review, which was the first step based on planktic and benthic forams across the Cretaceous–Paleogene mass extinction (1993). Two additional studies followed in rapid succession (1994, 1995). It was not until 2009 that I met my new collaborators, B. C. Jaiprakash and Napalla Reddy, and we began mass extinction studies in the Krishna-Godavari Basin based on deep drilling for many years of fun collaborations.

Deccan volcanic eruptions are known as Deccan Traps (from the Swedish word *Trapp*, or "stairs") that consist of terraced flood basalts, which are hot lava flows cooled into hard rock called basalt. These flood basalts form one of the largest volcanic provinces on Earth, originally covering an area of 1.5 million kilometers squared (0.6 million square miles) and spreading across half of modern India. (To give you a sense of scale, the state of Alaska is 1.7 million kilometers squared.) Over the past sixty-six million years, the Deccan Volcanic Province (DVP) had been reduced by erosion to about 500,000 kilometers squared (200,000 square miles), which is about equal to the combined size of Arizona and

Utah or the size of France. In today's India, the Deccan Traps still consist of vast mountains horizontally layered with basalt flows that stands like a 3,400-meter-tall wedding cake centered near Pune. These ancient lava mountains are known as part of the Western Ghats Mountain range that parallels the west coast of India.

The Deccan Traps contain the longest lava flows known in the world, which originated from eruption centers near Pune and traversed India over 1,000 kilometers to the east and into the Bay of Bengal. The conduits for these massive hot lava flows are believed to have been the channels through which the mighty Godavari and Krishna Rivers run. These parallel waterways, deeply carved into rocks like the Grand Canyon, originated near the mountains of Pune and drained rain from the Deccan mountains into the Krishna and Godavari Basins and the Bay of Bengal. Imagine, these two grand river systems turning into giant lava rivers speeding across the country. On top of the lava a cool, crusted surface formed, which trapped the heat underneath, creating lava tunnels. Similar lava tunnels on a much smaller scale are known all over India; many of them have carved monuments, most famous of all are the Ajanta and Ellora Caves in the Aurangabad district of the Maharashtra State.

Indian, British, and French research teams had been studying this region for decades. Most notably, in 1985, Vincent Courtillot's French team, with Indian collaborators, began logging, mapping, and dating the Deccan Traps. They found that more than 90 percent of the eruption volume occurred over fewer than 800,000 years, with the most rapid eruption rate and volume near the end of the Cretaceous. But evidence for the mass extinction remained elusive in this immense stack of basalt flows and Courtillot was obsessed with finding it.

Identifying the mass extinction required marine sediments with planktic forams, which were the only Fossil Group that suffered near total mass extinction across the Cretaceous–Paleogene boundary (KPB). In the absence of marine sediments, finding this mass extinction in the Deccan Traps remained a dream.

This dilemma was the subject of some friendly banter Vincent and I exchanged at conferences. When we bumped into each other he would routinely ask me: "Gerta, when are you coming to India to find the mass

160 THE LAST EXTINCTION

extinction boundary?" I would answer, "Whenever you find marine rocks between lava flows, I'll be there." This exchange went on for two decades, until one day, it was no longer a joke.

On the second Wednesday in October 2005, just after I'd finished teaching my three-hour class "Evolution and Catastrophes," my phone rang, and a voice exclaimed "Gerta! When are you coming to India to tell us where the mass extinction is in the Deccan Traps?"

The speaker needed no introduction. It was Vincent. His charisma transcended the wire. I noticed something different this time. His voice was as jubilant as if he just won the Nobel Prize.

"I have the locality for you," he nearly shouted into the phone. "It's in a quarry near the city of Rajahmundry, where marine sediments are between the longest lava flows across India."

This was the news I'd been waiting for. I was ready to start work on the Deccan Traps, and this Rajahmundry locality would be the perfect start. My mind raced ahead. I did a quick search on my computer. Rajahmundry was a mid-sized city along the Godavari River on India's east coast where the longest Deccan lava flows had traveled a distance over 1,000 kilometers and beyond into the Bay of Bengal. The vast size of the volcanic province boggled my mind.

"So, you're in?" Vincent asked. "My team will go there in early November. Can you join them?"

I was in, but the timing was bad because my teaching responsibilities continued until December 21. Fortunately, Thierry could join the French team in Rajahmundry in November and sample the marine sediments between the longest lava flows and ship forams to me at Princeton. I was skeptical that he would find forams. Rajahmundry was a beach area at the time of the mass extinction and few planktic forams existed in such shallow environments.

When I received the samples in the middle of December, I processed them immediately and searched for forams. As I expected, there were none in the first few samples, so I quickened the pace. Suddenly, some white specks appeared, and I zoomed in from 60x to 100x magnification. My jaw dropped as I recognized tiny forams. These must have been the first new species that evolved a few thousand years after the mass

TIME TO BE BOLD 161

extinction. Was this real? Could we have discovered the end-Cretaceous (KPB) mass extinction between Deccan lava flows on the very first try, let alone between the longest known lava flows? Cautiously, I looked again, adjusting for larger magnification and clarity. Yes! The newly evolved foram species were there, including the characteristic *Parvularugoglobigerina eugubina* that spanned the high-stress interval following the mass extinction. This meant the mass extinction occurred during the time of the massive lava eruptions that resulted in the longest lava flows across India. This was the Grand Slam!

I called Thierry and told him the good news: I had discovered the tiny, early Paleocene species that evolved after the mass extinction. Predictably, he was his usual skeptical self.

"Thierry," I prodded, "If I'm right about the forams, this is a big deal! We've got to go back to India for more samples. We need to be certain this is the end-Cretaceous mass extinction." After a brief pause, Thierry replied, "I'll arrange for the Indian guide to meet us in January." I thought I detected the hint of a smile in his voice.

On our first day in Rajahmundry in January 2006, Thierry and I walked along the sacred Godavari River with its giant temples and holy bathing sites. Throngs of people were celebrating the Hindu festival of Thaipusam, which was held on the brightest moon in January to celebrate the ancient Hindu god, Murugan. Men proved their loyalty, faith, and penance by body piercing with long metal rods and hooks. We stopped to watch a group of young initiates, naked down to a loin cloth, being prepared for their first piercing by an older man who placed a slice of lemon in each boy's mouth. He told them to have courage. He then pierced the cheeks of each one with a three-foot-long metal rod and told them to dance. It looked gruesome but no blood flowed from the wounds, and the men danced as though in a trance, seemingly oblivious to pain. Women participated in separate rituals with small thin rods piercing their cheeks and tongues.

With this annual ritual, the faithful hoped to be granted courage, wealth, and wisdom. They had the courage, but would they ever gain the wealth and wisdom if they lacked financial resources to gain an education? Perhaps this was the way to give that underclass hope that Lord

162 THE LAST EXTINCTION

Murugan would grant it. Deep down I felt it wasn't all that different from growing up in poverty in most countries.

The next day, we visited the city's active quarries, which revealed gaping holes in the earth that pockmarked an otherwise verdant agricultural landscape. We heard the quarry before we saw it. Above the mind-numbing rumble and roar of trucks climbing up and down the steep quarry roads, dynamite explosions roared like thunder. Thierry and I walked to the farthest corner of the quarry, away from the dangers of the exploding rocks. We climbed over a pile of large basalt blocks to reach the vertical wall, formed by layer after layer of what were once hot lava flows that cooled into hard basalt rock. We spotted what appeared to be irregular seams between the basalt flows and scrambled up the wall for a closer look. All basalt flows shared the same dark gray, finely crystalline appearance and were separated by short breaks in eruptions. And then the eruptions stopped for a long time, during which three to four meters of sediments deposited, known as intertrappeans—meaning in between the traps. Above these sediments, massive volcanic eruptions resumed.

"This is it!" I yelled exuberantly. I grabbed my hammer from the backpack and walked along the basalt ledge to clean off the surface and reach fresh rocks below that could preserve the thin-shelled forams. Thierry began drawing the rock column in his notebook, describing the different sediment layers, their thickness and different rock types. I collected the samples we would analyze in our laboratories to determine the age and depositional environment of the various sediment layers. Over the next few days, we sampled similar intertrappean sediment layers in three more quarries around Rajahmundry. In each locality, mudstone replaced by limestone marked an estuarine environment as the sea level transgressed for a short time. Then the sea level dropped, turning the seafloor into arid land, which was eventually covered by another series of four longest lava flows.

Back in my Princeton lab, I quickly processed the samples and searched for forams. Each of the four localities was similar with an assemblage of small early Paleocene species that evolved right after the mass extinction. I was ecstatic and ready to celebrate this amazing discovery. But I couldn't—not yet. To find the KPB mass extinction boundary so

TIME TO BE BOLD 163

easily at the first try in between Deccan basalts was too good to be true. Why had nobody else made this discovery? I read about the earlier studies on the Rajahmundry basalt flows and intertrappean sediments and found no studies on planktic forams. That explained it. No other microfossil group suffered near total species extinctions followed by immediate evolution of new species. Only planktic forams could readily identify the mass extinction.

Majda and I jubilantly celebrated my first KPB discovery in the Deccan Traps of Rajahmundry over dinner at Aqua Grill in Manhattan, New York. While waiting for our meals, Majda questioned me about every detail of my study, which I readily supplied until he was satisfied and happily agreed with my conclusions. The next day, I called my collaborators and friends to announce this historic discovery of the KPB mass extinction between the longest lava flows of the Rajahmundry quarries. They too were overjoyed. Vincent reminded me of his open invitation to visit the Geophysics Institute in Paris and we agreed on the summer of 2007. Majda visited the College de France, where he was an honorary member with the perk of a great apartment on Rue Jacob in the inner city. It was our most exciting and fun time in Paris. I spent my days at the Geophysics Institute working on the Rajahmundry study. In late afternoons Majda and I walked and talked through the gardens and parks of Paris followed by eating our way through the best restaurants. More difficult was finding time to talk to Vincent. He was a very busy man. On my arrival, I saw him for five minutes and he asked me to give a talk on the Rajahmundry study the following week. That talk was postponed week after week and never happened.

In October 2008, I presented our first discovery of the mass extinction between the longest lava flows in India at the Annual Geologic Society of America (GSA) Meeting. My talk was well received. In their *News Release*, GSA highlighted our discovery as the first evidence that Deccan volcanism was a major culprit in the end-Cretaceous mass extinction. The resulting publicity transcended science circles and reached mainstream audiences around the globe via sources such as *Discovery News, Science News, Nature,* and *The New York Times.* The author of the *Science News* article summed up the implications of the new findings well, writing, "The eruptions, which created the gigantic Deccan Traps

164 THE LAST EXTINCTION

lava beds of India, are now the prime suspect in the most famous and persistent paleontological murder mystery."[33]

In India, the discovery of the mass extinction in the Rajahmundry Deccan Traps received a lot of attention and triggered a stampede to find further evidence in the Rajahmundry quarries as well as elsewhere in the volcanic province. The excitement in India was palpable, when a new locality with KPB age intertrappean sediments was discovered at Jhilimili, north of Nagpur, as part of a Trans-India-Seaway. We couldn't wait to return.

The impact camp was unprepared for our discovery and remained relatively quiet because they had ignored Deccan volcanism and were unaware of its extent and importance. Jan Smit attempted to dismiss the findings by saying, "I don't believe they have said anything new here . . . when the meteorite hit, it would have put out more gas in vaporizing rock than the entire Deccan flow." Again, it appeared he had not fully understood the scientific findings: Vincent Courtillot and his colleagues reported the Deccan flows released some fifty to one hundred gigatons of sulfur dioxide and carbon dioxide—about ten times more than estimated from the asteroid strike at Chicxulub.

It was a great win to have discovered evidence of the mass extinction in the Deccan Traps, but we only had foram evidence from the earliest evolution after the mass extinction. But this was only a piece of the puzzle. We still lacked evidence of the mass extinction between the longest lava flows in the Rajahmundry quarries. My mind kept churning. We were so close—could we retrieve the extinction data in sediments below the lava flows in the quarries? It was worth taking the risk. I would pitch my drilling proposal to Leonard Johnson and Rich Lane, my NSF supporters. They believed the impact theory should be questioned like any science and contrary data revealed and considered. It was through their advocacy that the NSF had funded my drilling and research projects in Texas and Brazil. Perhaps they might be willing to fund my India drilling proposal.

In a fortuitous interview with *Discovery Channel*, I mentioned drilling as our next option in our Deccan research. To my surprise, they offered to fund the drilling in return for filming and publicity rights. I called Leonard and Rich at NSF to tell them the news. Stony silence greeted me. I had miscalculated the dynamics between government and media funding. Into this silence I said, "I've yet to agree, because I wanted to run it by you." Leonard and Rich urged me not to accept the funding. All too often, Lane explained, NSF funds the hard part of getting science projects off the ground. "When projects succeed, organizations like *National Geographic* and *Discovery* swoop in and skim the cream for a minor investment and we receive no credit."

"I can say no," I replied, "if NSF can fund the drilling. That's my only other option." I held my breath. "We will fund it," Leonard answered. His quick response floored me. I wasn't accustomed to getting my way so easily. Was it my lucky stars? Or was it my epiphany to follow my instincts and dare to think big? Barriers that had once blocked my path seemed to be falling away, and I had the sense I was now swimming with the current, enjoying a kind of effortless momentum in my career I'd never experienced before.

The Rajahmundry quarry drilling was set for December 2008. It was flawlessly arranged by a remarkable woman, Assistant Professor Nallamuthu Malarkodi from Bangalore University, who made all the arrangements with the quarry operator in Rajahmundry and hired a local drilling rig for the operation. In the morning the drilling began. Our small operation sat in a far corner of the quarry, away from the active mining and below the basalt flows. We would be drilling directly into the underlying Cretaceous beach sand that was running through my fingers. I put up a good face as the drilling began. Thierry, Malarkodi, and I watched the operation and waited for the drill to bring up the first cores. There was nothing but loose beach sand core after core. My worst fears were confirmed. Drilling doesn't work in loose sand, because when the core is withdrawn from the drillhole, the sand walls collapse, and every core afterward will bring up more loose sand; there is no end in sight. In consolidated sediments, a drill can go deeper and deeper. When I first saw the sand emerge from the drill, I knew it was total loss. "Shit," I muttered.

Thierry walked away and ran his hands through his hair. Malarkodi looked stricken. I sent the next drill down, thinking if it was only tens of meters of sand, we might be able to scoop it out. Again, we waited.

166 THE LAST EXTINCTION

Again, the drill regurgitated sand that fell apart in our fingers. It was time to pull the plug and save the money for a more promising project. I called off the drilling, thanked the operator, and paid for the day, plus a bonus day for trying.

It was a setback, but a minor one. I already had a better solution in the crosshairs—studying existing drilling cores taken in the Krishna-Godavari Basin by India's Oil and Natural Gas Corporation (ONGC). The problem was no one believed it was possible to get access. I had already made a few discreet inquiries about how open ONGC would be to academic collaboration, and the answer from my Indian colleagues varied from NO to NEVER. But I was not ready to give up without trying. NO was not an option.

13

"OOHING *and* AAHING OVER ROCKS"

IN THE 1998, A DANISH SCIENTIST NAMED HANS JORGEN Hansen took a plane from Miami to Copenhagen carrying precious cargo. In the large pockets of his overcoat, he carried a dozen crocodile eggs, wrapped in stockings for warmth. Hansen had come to Florida to collect these eggs for experimental studies into how dinosaurs died out during the Deccan volcanic eruptions sixty-six million years ago. The reason he had decided to study eggs was because thousands of unhatched, fossilized dinosaur eggs with unusually thick shells had been found throughout India, dating to the period around the Deccan eruptions. *Why*, Hansen and many other paleontologists wondered, *were there so many unhatched eggs?* The key to understanding what happened might be the crocodile eggs he was carrying with him on the plane. But first they needed to be kept warm. So, he cuddled them against his warm belly, like a mother cradling her newborn baby.

Back in his laboratory at the Geological Institute of the University of Copenhagen, Hansen placed the crocodile eggs in an incubator set to a temperature between 28 and 32°C to let the embryos grow and mature. After they had acclimated to the new environment, Hansen exposed each egg to mercury, chloride, fluoride, and sulfide to mimic the toxic environment of Deccan volcanism. Then he monitored them, periodically increasing the injection of poisons to higher levels for some of his eggs.

168 THE LAST EXTINCTION

After eighty to one hundred days of incubation, most embryos had died, leaving a rotten smell that pervaded the lab and leaked into the corridors and offices of his colleagues.

Upon dissection of the eggshells, Hansen learned a surprising fact. Exposure to the toxic volcanic elements had thickened the eggshells. This meant that the baby crocodiles that had survived the toxicity were unable to poke through the shells to freedom. He concluded that the abundant unhatched dinosaur eggs in India's thousands of nests had died for the same reason: Toxins released by intense and prolonged volcanic activity had thickened the shells, preventing the dinosaur babies from poking through and so they died.

The ancestors of crocodiles, of course, survived the KPB mass extinction. But there was still a lesson from his experiment, however crude. Hansen showed one facet of what the environmental conditions created by the massive volcanic eruptions at the Deccan Traps could do to the egg of the terrestrial dinosaurs' closest living relative. These results were like what we've seen in the fossilized remains of thousands of thickened dinosaur eggshells that never hatched in Deccan's volcanic environment.

Life can be both stubbornly resilient and surprisingly vulnerable. But even the hardiest of species can struggle and perish if the complex ecosystem to which it has specifically evolved is thrown out of balance. A food source might disappear. A new competitor species might emerge. Rapid changes in temperature might make the area in which a plant or animal lives uninhabitable. Impactors believe 70 percent of species died due to a shock event that rendered the world suddenly uninhabitable for them. But this was always an unusual hypothesis, because the other four mass extinctions—and many other smaller extinction events—are believed to have occurred gradually, due to changing environmental conditions. The Permian, or third mass extinction, was the most devastating the world has seen, resulting in the estimated deaths of 96 percent of species. Its cause was massive volcanic eruptions in Siberia 250 million years ago. There is also the inconvenient fact that abundant evidence from the fossil record leading up to the KPB shows a gradual decline in species occurring over a long period of time predating any proposed date for the Chicxulub impact, even the one favored by impactors. If species are

"OOHING AND AAHING OVER ROCKS" 169

dying over a long period of time, and then an asteroid strike causes a brief spike in extinction rates, does it make sense to say the asteroid was responsible for the entire extinction event?

Based on the evidence, it was reasonable to conclude that the fifth mass extinction was a gradual process, and that the cataclysmic volcanic activity that occurred during these millennia of species decline was responsible for it. But accepting this premise only led to many more questions—not the least being *How specifically did the volcanic activity lead to species death?* Hansen's study was just one of the many attempts to model the impact of the volcanoes on life. But there have been many others.

But all these theories depended on a better understanding of the eruptions themselves. How did they unfold? Did they happen continuously, or were they intermittent? How violent were they? And, of course, the million-dollar question: When exactly did they occur and did their occurrence coincide with the species deaths that occurred up to the KPB? These were the questions my team and I were determined to answer.

After our drilling failure in the Rajahmundry quarry, my team retreated to the quarry operator's office in a shelter perched on a high level of the quarry. It was a modest building, more of a lean-to with a roof that jutted out above the back wall of basalt. The front wall consisted of glass doors that provided a panoramic view over the quarry. The doors could be closed if it was windy or dusty, but that day they were wide open. We sat in plastic chairs around a coffee table, sipping hot tea and Coca-Cola in disconsolate silence. There wasn't much to say. Our mission had failed. I stared out across the quarry to the agricultural fields beyond. Sunlight sifting through clouds created a dappled effect on the landscape. The view was stunning, but the atmosphere in the room was dismal. It was time to regroup.

"Let's talk about our other options," I suggested. My colleagues exchanged uncertain glances.

"Perhaps we should go to Meghalaya," I said. Meghalaya has the most famous mass extinction sequence (KPB) in India, with one of the highest iridium anomalies in the world at twelve parts per billion (ppb). Located in northeastern India, Meghalaya is sandwiched between the Indian state of Assam to the north and Bangladesh to the south. At eight hundred

170 THE LAST EXTINCTION

kilometers from the fringe of the Deccan Volcanic Province, Meghalaya yielded a record of the toxic effects on life and the environment.

"I'm afraid that's impossible," replied Malarkodi. "The KPB sequence there is exposed along the border with Bangladesh, where kidnappings for ransom are common. The government permits no foreigners there."

I recalled two micropaleontologists, Rahul Garg and Vandana Prasad, from the Birbal Sahni Institute in Lucknow, Uttar Pradesh, who reported the mass extinction in Meghalaya.[34] I made a mental note to reach out to them with an offer of collaboration for access to samples, and a lecture during my next visit to India. This was the type of personal politics I was good at.

Finally, I described my plan to study the ONGC's sample cores. My colleagues were skeptical to say the least. I told them that to succeed where others had failed, I planned to launch an all-out charm offensive to gain entry to the ONGC. This would require a better understanding of Indian culture and the interaction between social and professional strata in the country. I needed to learn the different protocols required by different echelons of Indian society and how to address individuals based on their rank and position.

I spent the next ten months learning as much as I could about these complex social dynamics. One fact I picked up very early on was that Indian academics and the engineers at the ONGC were on very poor terms. Many of the scientists believed the ONGC was arrogant and excluded them and, as I soon found out from ONGC insiders, the feeling was mutual. This mutual animosity added another level to the complexity of my task.

When I was finally ready to make my approach, I called Sunil Bajpai, Vincent's fixer. After telling him I wanted to study the ONGC's core samples, Sunil immediately told me it was a waste of my time. The ONGC, he patiently explained, had never responded to academic inquiries whether from Indians or foreigners. "That doesn't bother me," I replied. "All I need from you is to find an ONGC insider to teach me the ropes: Who were the top chiefs, how should I formally address them, and what was their contact information? I want advice from an ONGC insider about what research information is crucial to ONGC, and how I might be able

"OOHING AND AAHING OVER ROCKS" 171

to help in that regard. Last, but not least, I planned to invite myself to ONGC's research campus in Dehradun to give a talk on the Cretaceous-Paleogene boundary in Rajahmundry in early February 2008. I want to know who to approach to make this happen."

Sunil rose to the challenge. He knew just the right man to help me. His name was Dr. DSN Raju, a retired foram specialist who spent all his life working for ONGC at the Dehradun Research campus and had great inside knowledge of its politics and how to get things done. DSN Raju, Sunil, and I discussed ONGC's inside operation, which was highly complex with branches throughout India, but the heart of it was the research campus of 2,000 people in Dehradun in the foothills of the Himalayan mountains. The most critical aspect of drilling for oil and gas is to know the age of the rocks as you drill deeper—that's where age dating based on planktic forams comes in handy. Here I had something to offer. I could provide more precise age dating of their cores and improve the accuracy of their drilling and potential oil and gas recovery.

DSN Raju also confirmed the tensions between the academics and ONGC employees. Indian academics, he told me, were notoriously arrogant and hated the oil guys. And the oil guys returned the sentiment. Knowing such tension existed between ONGC and the scientific community helped me craft my approach, which would be extremely respectful, friendly, and couched in the spirit of mutual collaboration between my team and ONGC scientists. I wanted to make it clear that I was not just asking for access to the cores, but that I was offering to work with their scientists in a way that would benefit us both and advance their reputation within the science world as well.

With DSN Raju's assistance, I sent my "self-invitation" to give a talk in early February 2008 at the Dehradun ONGC campus that would present the new Rajahmundry KPB findings and explain how this discovery might aid ONGC's deep drilling in the Krishna-Godavari (K-G) basin. To my amazement, my offer to give a talk was immediately accepted and arrangements were made for the day's event.

At the same time, I sent a letter to Dr. O. P. Pandey, director of ONGC in New Delhi, via FedEx and email. "To the Esteemed Dr. O. P. Pandey," I began, then introduced myself briefly and my scientific interests in India's

172 THE LAST EXTINCTION

Deccan Traps regarding the mass extinction. I explained the important discovery of the K–P mass extinction in the Rajahmundry quarries and how these findings could significantly improve age dating of the ONGC deep cores in the K-G basin. I emphasized collaboration with their scientists could be mutually beneficial to both sides. My major goal, I told him, was the recovery of the K–P mass extinction in the Deccan Traps, which would challenge the dominant meteorite impact theory. It was now clear the Deccan volcanism in India played a major role in this mass extinction, and ONGC held the key to solving this problem for the good of science. I ended with my offer of the public lecture to ONGC Dehradun for early February, to demonstrate my seriousness.

Everything I wrote was true but presented in a way to maximize the possibility of collaboration. Oil companies are usually seen as users and exploiters and not given respect by scientists. I presented our potential collaboration as ONGC's opportunity to play a role—a big one—in cutting-edge scientific research that could yield major international publicity. The good kind.

Dr. Pandey replied to my letter with great interest and warmth. Apparently, I'd hit the right notes and was on the right track. ONGC had cracked its doors open to listen and I had to make it worthwhile for this giant corporation to go the next step. Everything now depended on my February 2008 visit, and whether I could succeed in persuading the top brass to cooperate in this science adventure.

My research and learning about Indian culture had been intense and worth it. I was now confident of navigating my way through ONGC's complex hierarchical structure. It was the kind of politicking I never expected but enjoyed the inside peak into Indian culture. The hard part, however, was still to come.

Majda joyfully followed my team's successful discovery of the KPB mass extinction between the longest Deccan lava flows in Rajahmundry and kept his math friends abreast of the news with pride and praise. Most of all, he was happy for me to have finally left the Chicxulub impact controversy and the animosity that had followed me uninterrupted for twenty years. He believed I would be free now to pursue my science without the interference of the impactors. I didn't dare spoil his vision,

"OOHING AND AAHING OVER ROCKS" 173

but I knew he was dreaming. Though my presentation at Nice had undermined the core scientific arguments made by the impactors, the theory continued to thrive in the public consciousness and remained deeply embedded in scientific literature and institutions.

The idea that life all over the planet could be suddenly extinguished by a rogue asteroid thrived in apocalyptic Hollywood movies. What's more, multiple generations of students had been taught this theory as if it were established fact, and few understood the nuances of the arguments and data that contradicted it. I knew it would take years—perhaps decades—for the more complicated and accurate story of what really happened around the fifth mass extinction to be fully understand. I also knew that people who had dedicated their careers to the impact theory would not easily give it up. Nevertheless, the first couple of years spent conducting research in India were calm because the impactors were still in total denial that Deccan volcanism had anything to do with the mass extinction.

When I told Majda I was planning an important two-week trip to India in February 2008, he was alarmed. He didn't want to be left on his own worrying about my getting sick, as I did on my last trip. He wanted to be there to take care of me and I appreciated his concerns. But Majda, the mathematician, had no use for geology field trip adventures, and even less if it involved intensive politicking. I would rather be on my own. There would only be a few days of fun fieldwork and sightseeing in the state of Gujarat. After that, a lecture at the Birbal Institute of Paleosciences in Lucknow, the capital of Uttar Pradesh, where I had successfully arranged collaboration and splitting samples of the Meghalaya KPB section. Then an overnight sleeper train to Roorkee, and onward to Dehradun's ONGC headquarter for the all-important day-long meetings with scientists and officials from the bottom to the top chief—and last my lecture. This was not Majda's idea of fun, and it would not be easy to manage his expectations—in addition to navigating the high stakes visit to ONGC.

Sensing my reluctance, Majda brought it up over coffee one afternoon at our favorite Manhattan café. "G, you don't want me to come?" he asked, as the waitress delivered our cappuccinos and cake.

174 THE LAST EXTINCTION

"I'd love to have you," I replied, "but this is going to be a very hard trip of politicking and one-star hotels. You won't be happy."

"I can tell already you have too much planned," he replied. "That's why you need me to look after you. I can adjust to one-star hotels for a while."

I stifled a laugh. He most certainly could not, but maybe we could add a one-week vacation after my lecture at the ONGC and travel in luxury, I suggested. Majda accepted immediately. "Fantastic, G! And I'll do my math work while you are in the field."

I blew on my cappuccino. Majda doing math work from the field in India? That would be a sight to behold.

Majda and I flew into Mumbai, where we met Thierry and Sunil. Then we all took a flight to Bhuj in the westernmost Kutch district of Gujarat, where the airport is just a stone's throw from the Pakistan border. Within minutes we dropped off our bags at the Dollar Hotel and made our way into Bhuj for lunch, which got us stuck in the holy cow traffic. Bhuj is India's city with more holy cows than people. White gray, emaciated, placid cows standing or taking a nap in the middle of the roads, munching food from vendors, or scavenging in the piles of refuse. Nobody bothers them; the holy cows always have the right of way.

We collected Deccan volcanic rocks the next day along a railroad cut in the desert, while Majda walked for hours up and down the tracks deep in thought and scribbling on a notepad. He did what he promised: do math while we were, in his words "oohing and aahing over rocks."

We took a trip to the Rann of Kutch, the 30,000 kilometer squared expanse of salt marsh that borders India and Pakistan. There, a field of blinding, snow-white salt stretches to the far horizon—an awesome sight to behold. The sparse native population lives on elevated wooden platforms at the edge of the salt expanse, along with their goats and chickens, and yet everything is so orderly and clean you could eat off the salt floor. Stepping inside their domed, perfectly circular homes took my breath away. A kaleidoscope of colors and symbols were painted on the walls and ceiling, which were further decorated with beautiful quilts. Each house was an artist's palace that contrasted with the blinding

"OOHING AND AAHING OVER ROCKS" 175

white outside world. I couldn't resist buying some quilts to keep alive my memory of these people's artistic imagination gone wild in a world of white salt.

After the solitude of the Rann, the city of Lucknow in Uttar Pradesh was like another planet with its constant rumble of traffic, hawking vendors, and the poor camped on the sides of the city streets. This was our next stop at the Birbal Sahni Institute, where we stayed at their guesthouse We arrived on Saturday, and by Sunday night I was ill with diarrhea, high fever, and delirium. This was Majda's worst nightmare. Scared, he knocked on the door of the other guests and asked for help finding a doctor. Within twenty-five minutes, a doctor arrived and diagnosed too much water loss due to diarrhea. He left me packages of replacement electrolytes to dissolve in water that I was instructed to constantly drink. By noon the next day I had recovered enough to give my lecture. In the afternoon, I discussed our collaboration on the Meghalaya KPB sequence with Rahul and Vandana. The next day, we split their samples for my team.

By midnight, we made it to Lucknow's train station with Sunil to take the overnight sleeper train to Roorkee, in the state of Uttarakhand, where he is a professor at the Indian Institute of Technology (IIT), and we stayed at the guesthouse.

British-built railway stations in large, important Indian cities are show-stoppers. At night, they look like castles turned into wedding cakes lit with multiple colors that make your jaws drop in awe. Lucknow's station was one of them. Once inside the train station, we entered a different world where the holy cows mingled with the throng of travelers but also had freedom to use the rail tracks. Majda and I watched, mesmerized, trying to comprehend how this system could work happily. We had Sunil help us navigate to the right train and our sleeper cabin.

Home to over 2,000 professionals, the ONGC Dehradun campus was expansive and heavily guarded. Our car approached the first of several gates, where armed guards with dogs asked us to exit the vehicle to have our papers checked and the car searched for bombs. Sunil was not permitted to enter the ONGC campus. Once we'd finally passed through all the

176 THE LAST EXTINCTION

security checkpoints, we were taken to the main building. The campus was immaculate, its modern buildings surrounded by landscaping, the grounds studded with brilliant flowers and fruit trees.

A group of scientists greeted us at the entry. Majda was led away to an office to work on math. It was now 7 a.m. and I began a series of meetings with scientists chosen by ONGC to potentially collaborate with my team. Rather than grilling me on the specifics of my research project, the ONGC employees spoke about their research. I realized these were interviews and they were auditioning to work with me. Unfortunately, none had useful skills for my KPB project or any knowledge of the cores. This was going to be a major problem. I needed collaborators who could contribute to the project—if not with science, then with inside access to ONGC core data. I found out no scientists at Dehradun had this knowledge or access to the cores because this was the domain of the Chennai Regional Branch in the south of India.

My last interview was with ONGC's director of exploration, Dr. P. K. Bhowmick, a soft-spoken, impressive man with a kindly face. When I entered his office, he got up, smiled broadly, walked toward me, and offered a firm handshake: "I'm very much looking forward to this meeting," he said warmly. "Never before have we met a well-known academic scientist that treated us with such respect." I smiled and returned the compliment, happy to have done something right.

We quickly started in on science. He asked forward-looking, incisive questions that revealed he had done his homework regarding the KPB project I proposed. Our science discussion was freewheeling and exciting, covering topics from the Deccan Traps to oil recovery, and he clearly knew everything about past and future drilling in the Krishna-Godavari basin. I felt sky-high: everything seemed to work out at the top level and the project was already approved verbally.

When our meeting concluded, he offered to walk me to my lecture. As we traversed the quiet hallway, I asked him why nobody had been permitted to study the cores before. I needed the answer straight from the top brass because my Indian academic collaborators expected to be part of my team, and I had to tell them the bad news: They would be excluded from ONGC core studies.

"Indian academics don't show us respect," he replied. "They want our cores but don't offer anything in return. They don't care to collaborate with our scientists, and they don't share their data with us."

I'm sorry to hear that, I replied. Both sides could benefit.

"You are very different," he continued. "You treated us with respect. You proposed collaboration and sharing of data. And you didn't ask for anything in return, other than core samples. We are happy to work with you."

My heart leapt with excitement and joy. I had succeeded with ONGC collaboration.

Thank you, I replied. I'm delighted to work with you and your scientists. I know this project will be a great success for science and will bring you excellent publicity for the generous sharing of your core samples."

Grinning from ear to ear, he stopped in front of the lecture room and shook my hand, expressing his pleasure to have met me and that he was looking forward to this collaboration.

This interview had gone so well; I had passed the politicking test at the top level; the door was open for my collaboration and already approved. Now it was up to me to manage the science collaboration upon which the success of this project depended. It would be the most difficult part and prone to insider fighting for a seat at the collaboration table, which was seen as a step to promotion. There would be hard fighting between Dehradun and the Chennai drilling project. Who would be chosen as my collaborators? Who would control the project from ONGC's side?

My lecture was in a large boardroom of about thirty people—all top officials sitting around an oval table, with Dr. P. K. Bhowmick at the top, who introduced my lecture. Another thirty people sat around the fringes of the boardroom, including Majda. Tea and cookies were served. Only Dr. P. K. Bhowmick asked questions during and after the lecture. Then all filed out and disappeared. Majda and I were taken to the afternoon lunch buffet with a crowd of ONGC scientists. None had a clue who I was, apart from being the wife of a scientist.

The men clustered around Majda and questioned him. When he revealed he was a mathematician, they said only statisticians were useful

178 THE LAST EXTINCTION

in mathematics, which grossly insulted Majda. Soon I extracted him from the crowd before Majda's insult escalated.

I recognized just one person in the lunch crowd, DSN Raju, the retired ONGC foram specialist who taught me how to navigate through ONGC's top brass. I thanked him for his fantastic help and asked for further advise on how to deal with inside politics of the Chennai Branch of scientists in the project. He smiled happily and said we'd talk about it later, as he furtively looked around the room to see whether anyone had listened to my query.

In the ONGC limo, I leaned back into the seat, rested my head on the headrest, and closed my eyes, happy but exhausted. It had all been worth it.

Later that spring, I received an email from the directors O. P. Pandey and P. K. Bhowmick to submit a short list of proposed core sites to be studied in my project. Napalla Reddy and B. C. Jaiprakash (known as JP) proposed to analyze the best cores in the Krishna-Godavari (K-G) Basin along with new drilling, both onshore and offshore. That's when I discovered Reddy and JP were an incredible team of scientists with expertise in benthic and planktic forams, sea level and climate changes, and much more. Reddy oversaw related science projects in the Chennai labs and JP concentrated on drilling for oil and gas recovery. They deeply impressed me with their in-depth knowledge of dozens of cores and numerous ongoing drilling operations. This was my dream team collaboration I believed would succeed solving the KPB mass extinction in the Deccan Traps.

I had some prior knowledge of these scientists in 1993 when I was asked to review a paper on Deccan volcanism in the Krishna-Godavari Basin by the Geologic Society of India by authors PC Jaiprakash and DSN Raju. That review was my first indication that Deccan volcanism and the mass extinction were closely related but the ONGC cores were inaccessible. I had kept that old review to someday succeed—that day came in February 2008, the day I met Jaiprakash and Reddy at the New Delhi airport.

By the time I got through customs in New Delhi, it was well past 11 p.m. I had not met Jaiprakash or Reddy before and had no idea what they looked like. I scanned the crowd of waving people and spotted two men hidden behind elaborate bouquets of yellow and red roses. I assumed they were meeting their wives or girlfriends—why else such large bunches of flowers? But one man waved a paper with my name on it. As we greeted each other, he introduced himself and pressed the roses into my arms as a welcome gift. It was a pleasant surprise. We took a taxi to the Luxury New Delhi Old Grand Hotel for me to check in for the night. ONGC had rolled out the red carpet and reserved a suite of three large, elegant rooms. At 6 a.m., in the New Delhi fog of February that grounds all airplanes, we boarded the train that took eight hours to Dehradun. ONGC had booked my train ticket as first class but for their Chennai employees it was second class. After the train departed, I lugged my suitcase into second class and joined Jaiprakash and Reddy.

On this eight-hour train ride we discussed the KPB project, and I learned about their work and struggles to do science inside the ONGC. The two were close friends who complemented each other, and both were from the south of India. Reddy was the serious, quiet, academic type, always conceiving improvements to the KPB project. He proposed to plan, organize, and track down ONGC data of the dozen cores we chose to analyze for benthic and planktic forams. JP was tall, outgoing, talkative, and always smiling. He spent most of the time in the field drilling new localities in search of oil and gas in the K-G Basin. For our KPB project, we added samples recovered from half a dozen new drilling localities. Together they devised endless ways to tweak the project to ensure its success. It was JP and Reddy who advised me how to navigate the political mine field to achieve approval for our various plans.

After our fantastic planning session during the train ride, Reddy and JP cautiously informed me that the ONGC Chennai branch had approved their participation in my KPB project, as I requested, but under the condition that their normal day-to-day duties for ONGC continue. This threw a pall over my happy expectations for our collaboration. How could they realize even a fraction of their ideas for this project in addition to their full workload? They assured me they could integrate the KPB project into their everyday duties, but lab research would be limited. By the end of the train ride, I was convinced I couldn't have chosen better collaborators.

180 THE LAST EXTINCTION

After our visit to Dehradun, Reddy and I set out for Rajahmundry to collect samples from the ONGC core library, and JP headed to drill a new location in the K-G Basin. In Rajahmundry, Thierry Adatte and my graduate student Brian Gertsch awaited our arrival outside the core library for two days because they were not permitted entry without my presence.

The Rajahmundry core library was part of a large, heavily guarded compound. ONGC spared no expense in protecting these drill cores, which held evidence of oil and gas, volcanic rocks, and marine sediments. The building was huge, pristine, well lit, and air conditioned. Toward the entrance was a long table, and beyond it, row after row of meticulously kept shelves housing the cores. It was like a luxury hotel for cores, I thought. It had been a long road to get there, and my heart raced with the knowledge that the prize I'd long sought in Deccan volcanism was within reach.

Reddy, Thierry, Brian, and I gathered around a long table with our notebooks, plastic sample bags, Sharpies for labeling, and hammers for splitting rock samples. The librarian plucked the cores from the shelves and one after another laid them out on the table. Reddy had tracked down the ONGC drilling data for the dozen cores we planned to analyze. Each core we looked at had been drilled at localities that contained evidence of the two mega-pulsed eruption phases that resulted in the longest lava flows, which spread 1,000 kilometers across India. The first mega-pulse, we knew, ended at the KPB. The second began about 400,000 years after the mass extinction. The core samples showed a thick interval of sediments separating these mega-pulses just as we had observed in the Rajahmundry quarries.

But that's where the similarity with Rajahmundry ended, because these were the shallow surface eruptions from the Deccan lava flows that erupted at super speed through the hot lava tunnels, from the Deccan mountains to the Krishna and Godavari basins over 1,000 kilometers, and into the Bay of Bengal. Each of the two mega-pulses consisted of two to four of the longest lava flows separated by short "intertrappean" intervals of sediment deposition. We hoped to find the deeper water environment of the Krishna-Godavari Basin near the Bay of Bengal,

"OOHING AND AAHING OVER ROCKS" 181

where sediment deposition would be higher in quiet waters than in the Rajahmundry beach environment. In these sediments, we had the chance to observe the progress of the mass extinction after the eruption pulses. We systematically sampled each core sequence from the older Cretaceous to the younger Paleocene sediments.

Over the next four days we were the perfect team, working uninterrupted for eight hours a day from morning until closing time. As the days went by and the story unfolded in these layered sediments, our exhilaration and excitement skyrocketed. We were on the cusp of discovering the first direct evidence of the mass extinction directly tied to Deccan volcanism and the environmental catastrophe that killed off marine life. It was a dream almost too good to be true.

Eager to begin my lab analysis at Princeton, I decided to haul over 100 pounds of core samples back in my luggage, but ONGC protocol required they ship the core samples. I feared this might introduce needless delays at ONGC, with the added time of shipping, customs, and the possibility of a lost shipment. Plus, there was always the possibility that Dehradun's general lab manager would purposely delay or loose the samples we collected. Reddy found a way to get around ONGC rules by providing an official letter from ONGC describing the samples for scientific analysis. Still, there remained the problem of walking 100 pounds of Indian dirt through customs. How would I convince them? I assured Reddy I had experience taking precious cargo through customs and told him of the time I smuggled my pet tortoise, a protected species, through customs in Tunisia.

That adventure, I told him, occurred when I was doing fieldwork in Tunisia in 1998 with Thierry in an isolated area of the mountains near Elles. At one point during our work, a young boy came by on a donkey holding a small tortoise in his hand. "What are you doing with that?" I asked curiously. "My mother is going to make it my lunch," the boy replied. My heart immediately went out to this poor animal. Thinking of my pet tortoises at home, I couldn't let this one be eaten. I held out my hand, and the boy gave the tortoise to me. In return, I gave him a five-dollar bill, the smallest bill I had. I took care of the tortoise for the rest of the field trip, but by the time it came to an end, I had failed to

182 THE LAST EXTINCTION

find a safe home. So, I decided to smuggle it back into the US. Knowing Tunisia jailed anyone smuggling this protected species out of the country, I knew it was a big risk.

It was hot on the day we departed, but I went to the airport wearing my red and black Sierra parka, with the turtle nestled in its large outer pocket. When we got to security, I held my breath and walked confidently through the scanner. No alarms sounded. I was relieved—but too early. As I passed through, a customs official walked up and poked my bulging pocket with a three-foot-long rod. "What do you have in there?" he asked. *Oh shit, I'm going to jail,* I thought. Putting on my most sunny smile, I replied: "It's a tortoise, do you want to see it?" He peeked and asked, "Why do you want a tortoise?" "Because I love it," I replied smiling, raising my shoulders to my ears like *why else?* He shook his head. "Really?" he asked. As I nodded yes, he replied, "OK, you can go." That was how Elles joined my tortoise family.

It turned out getting the ONGC rock samples out of India and through US customs was easier than smuggling a tortoise out of Tunisia. As a geologist, nobody ever asks me why I am carrying rocks—that's just what we do. After I got back to my lab in Princeton, I immediately began processing the samples for forams and doing age and environmental analyses. I began with the mudstone samples from the core below the first major Deccan eruption pulse and longest lava flow. This was the interval we tried to core in the Rajahmundry quarry and got stuck in beach sand. We now had exactly the right cores from the Krishna-Godavari Basin located in the deepening marine environment toward the Bay of Bengal approximately sixty-five kilometers east of Rajahmundry's beach deposit.

The forams in this pre-extinction mudstone had high species diversity that rapidly decreased in abundance. Analysis showed that 50 percent of the species died out over just a few thousand years during the ramp-up to Deccan volcanism's extreme eruption rate and volume. The first extreme eruption sent lava flows from Pune 1,000 kilometers across India to Rajahmundry and into the Bay of Bengal. All but seven of the remaining species went extinct in the hours, days, or weeks it took for this violent and dramatic lava flow to traverse the continent. Three more of these longest lava flows followed, each over short time intervals of weeks,

The view of Mahabaleshwar Mountains in India shows step-like, or layer-cake type mountains of the Deccan Traps ("traps" is Swedish for step-like) mountains. As far as you can see these magnificent volcanic eruptions continued seemingly forever in horizontally stacked up layer upon layer. In this mountain world beauty, science and adventure met with perfect synchrony. For years we explored this world, and one day the answer stared me in the face: These are redbole clays, deposits from distant volcanic eruptions containing zircon crystals that could provide crucial age control. And there was more: ash from Deccan eruptions, carried across the globe over the course of one to one-and-a-half years and settling worldwide. These deposits could hold the key to our answers.

months, or years, ending life of all planktic forams, except one survivor: the disaster opportunist.

We now had evidence that the largest pulsed eruption phase in Deccan volcanism caused the end-Cretaceous mass extinction. Still, it was hard to believe how quickly and thoroughly this mass extinction ended life for planktic forams. What had happened? The foram shells that were left behind tell the story like scattered corpses on a battlefield.

The first half of the species that went extinct coincided with the increasingly higher rate of volcanic eruptions, decreased species populations, decreased species abundance, reduced shell size, and thinning shell walls. These were signs of high-stress environments caused by the ramp-up in the Deccan eruption rate, releasing increasing volumes of carbon dioxide, and nitric and sulfuric acids that resulted in increased temperature, toxicity, acid rain, and ocean acidification. These are the

184 THE LAST EXTINCTION

major killers of marine and terrestrial life during Deccan volcanic eruptions that led to the mass extinction.

Climate warming results from the greenhouse gas carbon dioxide (CO_2) from volcanic eruptions or fossil fuel burning, injected into the atmosphere where it traps heat and reflects it back to Earth. When climate warms 2°C or more, the threshold or tipping point for marine and terrestrial life is reached, causing mass extinctions. Acid rain is toxic to life whether on land or oceans because nitric and sulfuric acids combine with CO_2 in the atmosphere or seawater.

Ocean acidification results from the ocean's take up of approximately 30 percent atmospheric CO_2, which dissolves in seawater and turns into carbonic acid that decreases the ocean's pH. This process steals the carbonate ions needed by marine organisms, like forams, corals, crustaceans, mussels, and other shelly creatures, preventing them from fully growing their shells in the increasingly acidic waters. It's a creeping death, resulting in dwarfed and thin-walled shells, and finally extinction as ocean waters turn more and more acidic. In the K-G Basin of India, it took a few thousand years to kill off the first 50 percent of the planktic forams, but less than one hundred years for the remaining species as volcanic eruptions doubled or tripled the gas and volume released. These effects are not restricted to India but occurred on a global basis prior to the KPB mass extinction. Increasing ocean acidification leads to early extinctions throughout the food chain as larger animals feeding on calcareous plankton lose their livelihood.

As we know, the sole KPB planktic foram survivor was the disaster opportunist *Guembelitria cretacea*, which spawned environmentally tolerant species that dominated for the next 400,000 years, a time of cool climate and low Deccan eruptions. Afterward, the last major Deccan eruption pulse blew its top and sent another four lava rivers across India into the Krishna-Godavari Basin and Bay of Bengal. Just as before, the eruption center was in the Western Ghats near Pune and the hot lava rivers used the same lava tunnels of the Krishna and Godavari Rivers for transport across India. But that's where the similarity with the KPB ended. The difference in the environmental effects was like day and night. Whereas the end-Cretaceous eruption pulses caused the rapid

"OOHING AND AAHING OVER ROCKS" 185

mass extinction, the Paleocene eruption pulses caused no significant extinctions, species thrived, diversity increased, and climate warmed. The reason was simple: The Paleocene eruptions were well-spaced and permitted environmental recovery after each eruptions pulse.

Transport of lava such as we see in the Krishna and Godavari River channels, which run over 1,000 kilometers across India, are unique. But shorter lava tunnels are common in the Deccan Volcanic Province. Study of Hawaii's volcanic eruptions provide a current example of how lava tunnels are created. Hot lava rivers may flow into existing subterranean cavities or flow above ground in existing river channels. The lava river's outer edges cool and harden into a lava crust leaving the center as fast-moving lava river. The enclosing crust superheats the lava at temperatures exceeding 1,090 °C, which then melts down into the underlying rocks and creates a larger and deeper lava river that moves at high speed. This process continues if there is a supply of lava eruptions.

Now imagine the hot lava flows, emerging from the largest mega-eruptions in the Western Ghats searching for a way out. They find that escape in the Krishna and Godavari River gorges. The red-hot lava races through the canyons at high speed. The lava surface eventually cools and forms a crust, but the enclosed lava superheats and cuts down into the underlying rocks. This enlarges the lava river's size and transport capacity across India. The lava river perpetuates itself if the mega-eruptions continue in the Western Ghats. By the time the lava river reaches the lowland of Rajahmundry and the Krishna-Godavari (K-G) Basin, it splits and spreads out over the area, filling in existing depressions. This happened four times prior to the KPB mass extinction as the mega-eruptions continued in rapid successions.

During the early Paleocene, about 400,000 years after the mass extinction, another Deccan mega-eruption pulse exploded in the Western Ghats and repeated the same lava river journey four times, using the same lava tunnels. This stacked up the basalt flows in the same area. No significant species extinctions occurred at this time because the eruptions were widely spaced, which permitted species recovery.

186 THE LAST EXTINCTION

Massive volcanic eruptions split earth apart along a string of violent fissure explosions pouring out magma over hundreds to thousands of kilometers. This was the vision my graduate students and I conceived of Deccan volcanism in India and the death of the dinosaurs. Our idea of Iceland's Laki eruptions in 1783–1784, but a magnitude smaller compared with India. Similar fissure eruptions are well-documented in Iceland and illustrate the immense destruction caused by large eruptions and poisonous volcanic gases that led to widespread devastation and famine in Iceland and beyond.

Contemplating these vast and rapid volcanic eruptions as the cause of the mass extinction was mind-blowing. My dream of discovering the KPB mass extinction directly linked to the Deccan Traps had turned into reality even beyond my wildest imagination. The environmental disasters of this catastrophe revealed an astounding and groundbreaking new mass extinction scenario. It wasn't just an instantaneous event like the proposed asteroid impact, but rather a rapidly evolving unstoppable volcanic catastrophe.

According to the evidence we had found, the fifth extinction began with a ramp-up of increasingly larger eruptions in the Deccan Traps. These rapidly warmed the climate, acidified the air and oceans, and released toxic elements that killed off 50 percent of planktic forams over a short time. This is likely when the last of the dinosaurs—which were already in decline—were killed off all over the globe. It makes sense then, that in India the last dinosaur bones and nesting sites disappeared earlier, as proximity to the source of the increasing heat and toxic elements

"OOHING AND AAHING OVER ROCKS" 187

poisoned their food source, and thickened the eggshells of baby dinosaurs, preventing them from picking their way out. Shortly (in geological terms) after this ramp-up, the even larger mega-eruptions sent the mass extinction into sudden overdrive as the world's longest lava river raced 1,000 kilometers across India toward Rajahmundry's shoreline and spread out into the deepening sea across the K-G Basin and into the Bay of Bengal. This first mega-eruption killed off all but seven of the 50 percent of foram species that remained.

Three more of these mega-eruptions rapidly followed, each sending lava rivers across India. This left just one survivor species, the hardy disaster opportunist, *Guembelitria cretacea*. This species would give rise to a few environmentally adapted species within the first few thousand years after the mass extinction. But this species was one of the few that could survive the hostile conditions. The catastrophic events in northern India left oceans across the globe generally unfit for survival. It took 400,000 years in this high-stress, cool environment for more diverse but still small species to evolve. After that, India's Deccan volcanic plumbing once again blew the top, emitting another four mega-eruption pulses racing through the same lava tunnels 1,000 kilometers across India. But these mega-eruptions were widely spaced in time, permitting life to recover after each lava flow. Thereafter, Deccan volcanism ended its reign of terror, and the real recovery began.

It was time to share the Deccan mass extinction discovery. I called Dewey McLean, who had the most at stake because he originally proposed the Deccan Traps as the cause of the mass extinction before the impact theory was ever published, and he suffered tremendously from Alvarez's attacks for standing his ground. Dewey was thrilled to finally get his moment of vindication. He ended our conversation with his happy conclusion: "I always knew I was right, and you would eventually find the evidence." Chuck Officer was equally happy when I called, and he wistfully wished he could have participated in this joyful discovery. Vincent Courtillot was happy the Deccan Traps could now be directly linked to the mass extinction. He longed to continue studying the area. For him, the big question was still the same: Where in the Ghats were the big lava eruptions that caused the mass extinction? He was waiting

188 THE LAST EXTINCTION

for that discovery. Of course, we had made that discovery already in the Rajahmundry traps where the largest volcanic eruptions occurred and caused the mass extinction.

Finally, I called my NSF supporters, Leonard and Rich, to let them share in our monumental discovery, which they had so generously funded. We planned to announce the new findings at the 2009 annual AGU meeting in San Francisco. Capitalizing on our Deccan studies' international attention, NSF commissioned their artist to paint a panel depicting the death of the dinosaurs. Based on my advice, and what we know from the Deccan volcanic eruptions, I chose to illustrate the similar though much smaller Laki eruptions of Iceland from June 1783 through January 1984. At that time the Earth split open in fissures over hundreds of kilometers, spewing out clouds of poisonous gases into the air, and red-hot lava pouring over land lighting up the sky, while the great beasts writhed in poisoned agony. This is the type of volcanism that likely fueled the Deccan Traps, but thousands of times more deadly than the mass extinction.

Public recognition soon followed. In the fall of 2009, the NSF press release titled "Did Volcanoes Kill the Dinosaurs?" commissioned the image. By December, AGU followed with their own press release: "Volcanoes and Mass Extinctions." On January 9, 2009, the History Channel released the documentary First Apocalypse produced by Flight 33 Productions. It covered the various theories of the dinosaur mass extinction beginning with the impact theory and ending with the volcanism theory. Our Deccan volcanism scenario was picked up and reproduced in numerous media reports and magazine articles. We were riding a high tide of publicity I never thought possible.

14

TURNING *the* TIDE

I WAS OFF TO A GOOD MORNING. I'D ARRIVED AT MY SUNNY office, watered my plants, and sat down to finalize writing the results of my ONGC core study. Our research on the Deccan Traps was riding an unprecedented wave of success. My outlook held all the optimism of spring. In three days, it would be my sixty-fifth birthday. Majda had planned a big celebration at Aqua Grill in New York. In early June, I would treat two of my favorite sisters, Rosie and Helen, to a wild ten-day trip through the spectacular scenery of the Southwest. We planned to visit the Grand Canyon, Bryce Canyon, Monument Valley, and much more. It was my tradition to treat my sisters to an exotic vacation every few years. We shared the same adventurous spirit and had enjoyed adventures in Mexico, Guatemala, Morocco, and Egypt.

But the old headaches with the hard-core impactors persisted. Years had passed since crucial parts of their theory had been publicly disproven at Nice. Faced with our evidence, many in the field had moved on, abandoning the idea that an impact was the sole cause of mass extinctions. But the most fervent impactors stubbornly fought on. In particular, they were desperate to undermine the evidence my students had discovered at El Peñón in 2000 that showed that the spherule fallout from the impact predated the mass extinction by 200,000 years. In the years immediately following the discovery, we had been unable to finish our study of the

190 THE LAST EXTINCTION

site because Mexico's drug war had made the area far too dangerous for fieldwork. But by 2008, Thierry and I risked it all to revisit El Peñón to complete our research and we were ready for publication. Bringing our findings to the larger scientific community, however, would entail further delays.

Three times our research was rejected by different journals, and each time it was due to the intervention of the same two reviewers and collaborators: Jan Smit and Peter Schulte. It was not hard to figure out why. Our data clearly placed the Chicxulub impact well before the KP boundary, and thus undermined any theory that made an asteroid the sole cause of a sudden, catastrophic extinction event. They could not accept this, but they faced a problem: At El Peñón, the spherule layer and the KP boundary were separated by over twenty meters, including a major hiatus. If the asteroid caused the mass extinction, then how could the evidence of its impact be located so far from its supposed result. Rather than admit the obvious, they conjured up another ad hoc theory to explain away inconvenient facts. In this instance, they claimed the deeper spherule layer we had studied had been subjected to "liquefaction," and had sunk meters below the seafloor in soft sediments. Nothing like this had ever been proposed or observed before. Even as they put forth this implausible theory, they admitted they were unable to suggest any mechanism that could explain how it was possible, or why such a thing had never been seen elsewhere. Nevertheless, on such reasoning our research was deemed unsuitable for publication.

Three times I wrote letters to the editors of each journal pointing out the bias and erroneous reasoning of the reviewers and protested that their journal's decision not to publish did not take into consideration our revisions, critiques, and replies. Eventually, the editor of the *Journal of the Geological Society of London* replied, writing, "After reading your letter, [I] read the paper myself and found your data well presented." He accepted our paper and expedited its publication in 2009. We had finally succeeded in publishing our paper despite attempts to block it by the impact cabal.

Later that year, I put the final touches on the Deccan volcanism study. It was one of the most exciting and significant papers I had ever co-authored. We had already announced the initial results in the fall of 2009. Since then, my colleague JP drilled several new localities in the Krishna-Godavari Basin of India that widened and strengthened the scope of our investigation. During my visits, Reddy and JP took me on fun field trips to southern India where we explored several late Cretaceous outcrops for signs of environmental effects of early Deccan volcanism. They also introduced me to their cultural heritage in the town of Mamallapuram in Tamil Nadu, where we visited the unforgettable seventh-century Hindu monument known as Arjuna's Penance. Measuring forty-three by ninety-six feet, this giant slab of carved stone depicts South Indian life and myths. It was a wonder to behold. We bonded over these trips and became friends for life, continuing to work together on various projects, even after their retirement.

By 2010, our research team was ready to submit our paper presenting the comprehensive evidence we had gathered of the direct connection between the fifth mass extinction and the Deccan eruptions. There was much to look forward to. Then I received a call from a science journalist for Reuters named Kate Kelland. She wanted my response to the "news" that an asteroid had wiped out the dinosaurs.

"That's hardly news," I replied. "That theory has been around for thirty years."

"It's now official," she told me. Apparently, *Science* was about to publish an article signed by forty-one international scientists who reviewed the last thirty years of research and concluded a giant asteroid smashed into Earth bringing about the dinosaur extinction.

"Your Deccan volcanism research is dead," she said. "What do you have to say?"

"That it's nonsense. I can't say more until I see the article. Can you please email the embargoed copy to me?"

She hung up.

The journalist's "gotcha" tone and hostile attitude annoyed me, but I wasn't too concerned by the call. It sounded like another public relations stunt engineered by the impactors. The fact that there were forty-one co-authors was laughable and made the story seem totally fabricated. The average number of co-authors for a geoscience paper was five to six.

192 THE LAST EXTINCTION

Sometimes deep sea-drilling projects could reach two to three dozen authors, if there was a large crew. But forty-one co-authors of a standard science paper was unheard of. No credible geoscientists would fall for it.

I returned to my Deccan volcanism mass extinction study. It would be good to get this paper out as soon as possible to counter whatever nonsense the impactors were brewing.

Then the phone rang again. This time it was Paul Rincon, a reporter for BBC News. "Gerta, the dinosaur extinction linked to the crater is now confirmed by forty-one authors. This consensus is unshakable," he announced. "What do you think?"

And so it went, all day long. By late afternoon I was exhausted, worried, and had a splitting headache. I called Majda, who urged me to come home right away. When I stepped into the house, Majda pulled me into a firm hug. "I'm sorry, Liebling," he said as he held me. "You've been sandbagged!"

In the hours before the *Science* article published on March 5, the media chummed the waters, driving the hype that Chicxulub was the universally agreed upon culprit of the KPB mass extinction. When I finally got my hands on the paper, just in time for my birthday, I discovered that *Science* magazine had published the article not as a standard research paper, but as a "review." I was also surprised to see that the primary author was Peter Schulte, who, with Jan Smit, had attempted to bury our El Peñón research before. Half of the other authors I had never heard of, and half had never worked on the KPB mass extinction. Fifteen were hardcore believers in the impact theory. At best, half a dozen had enough expertise in the field to responsibly weigh in on the debate.

What was immediately clear was the article—which contained no new scientific findings—was a defensive response to evidence my team and I had presented at Nice in 2003, as well as our recent discoveries in India.[35] The third paragraph mentions "other interpretations of the K-P boundary mass extinction" based on "stratigraphic and micropaleontological data from the Gulf of Mexico and Chicxulub crater," but made no attempts to refute our data or conclusions. Instead, the article summarized the familiar mix of misrepresentations, discredited facts, and half-truths that had been made in support of the impact theory since

1990. There was a long section on the distribution of ejecta from the Chicxulub impact that was meant to establish the events correlation with the mass extinction. Much of the data presented in this section merely affirmed there was an impact at Chicxulub—a fact that no one contested. When forced to reckon with the fact that spherule layers from the impact were located tens of meters from the K–P boundary, these findings were explained as "high energy sediment transport" caused by "tsunamis and gravity flows" and "impact-related liquefaction."

The only novel arguments were attempts to trash the recent findings on Deccan volcanism, calling them inconsequential to the KPB mass extinction. In the absence of new findings, or any serious critique of our research, this was an unusual article for a professional journal to publish. *Science* publishes original research, short notes, and review articles. A review in a scientific journal usually includes a thorough reckoning of all the scholarship produced on a given subject, sometimes going back decades, and offers a conclusion as to what's true and accurate in light of the most recent research. Schulte had essentially summarized the arguments of one side of an ongoing debate and then attached a bunch of names to it to lend authority. There was no sense that this was an open matter of inquiry that should be settled through ongoing scientific research. It was an attempt to establish a settled narrative and to preempt further inquiry. It was reminiscent of the farcical loyalty oath exercise I had witnessed at Snowbird III—a statement of allegiance to a threatened theory.

This tendentious summary of the debate ignored or misrepresented decades of scientific work that my team and others had disproved their theory. It presented the impact theory as unimpeachable. The asteroid that created the Chicxulub crater was the true catalyst of the KPB mass extinction. Case closed.

It was galling to me to see that a journal dedicated to science had become a vehicle for propaganda. I could have accepted a decision to take a pro-impact position that at least engaged with the evidence. The editors of *Science* needed only to read their own journal to find it—my co-authors and I had already addressed and disproved the major points in previous papers as well as in the "comments/critiques" we'd written in reply to

194 THE LAST EXTINCTION

earlier papers arguing for the impact theory. They could even have asked me to comment. But all of that was ignored in this "review."My name was littered throughout the paper in negative, dismissive references to my publications.

During my deep annoyance as I read the article, I was struck by a thought: The impact camp had nothing new to say. In the seven years since the Nice EGU-AGU meeting, my team pioneered original research proving that Chicxulub didn't fit the KPB age, and we were now announcing new discoveries in the Deccan Traps. We had written dozens of papers during this time and had begun to receive attention in the science and popular press. In the past three years, the GSA, AGU, and EGU meetings had focused on Deccan volcanism, because we were discovering new things.

Meanwhile, the impact camp had little fresh evidence to report. And what "evidence" they found that the impact occurred at the same time as the mass extinction increasingly relied on familiar and spurious interpretations of the facts, or postulations of phenomena that no one had ever before witnessed, like massive tsunamis that don't disturb life on the ocean floor or spherules that conveniently sink through layers of sediment that accumulated over hundreds of thousands of years. These far-fetched ideas were driven by circular reasoning: *The impact caused the mass extinction therefore the impact is KPB age.* Facts must be twisted to suit the theory. With Nice, and the recent findings on Deccan volcanism, impactors had been caught on their heels, so they had to do something to stifle my team.

Then there was the personal angle in the attack. The *Science* review was written not just to counteract my team's success but to bury me. In Peter Schulte the impactors had found a willing instrument of destruction. Eight years earlier, Peter had been a visiting graduate student in my lab at Princeton University, where he worked for two summers (2001–2002). He'd made a strong, and unfavorable impression on me. I could still see him clearly in my mind—tall, blond, with a stiff, imperious bearing and a habit of eating food off other people's plates. He had been Wolfgang Stinnesbeck's graduate student at Karlsruhe University and was sent to my lab to learn the ropes of stratigraphy and age dating. That's

TURNING THE TIDE 195

what I taught Peter, and I also did the age analysis for his thesis for the KPB sections in northeastern Mexico. Peter was smart but not a team player. He joined my crowded lab already filled with my four graduate students. They didn't like Peter.

One day, all four of them appeared in my office with a litany of complaints, most seriously that Peter hogged my lab's newest computer and would not let them use it. When I asked Peter why, he explained that he had installed a new graphics program licensed to Karlsruhe University, and therefore he was obligated to prevent other students from using the computer, which he said would be illegal. It was an astonishing argument to justify hogging my new lab computer. I told Peter that he must remove the program and let my students use *their* computer or he could leave my lab. To my amazement, Peter packed up and left. Not long after, he joined the impactors and spread rumors that he fled my lab because I prevented him from joining the impact camp. The impactors loved having a defector from my lab, someone in their ranks who would do anything to get back at me. I didn't think enough of Peter's work to consider him a threat, but I was wrong. I'd once again underestimated the impact camp's talent for dissembling.

As publicity for the *Science* review snowballed, my detractors piled on. Peter Ward, a former friend turned impactor from the University of Washington, Seattle, sent me an email that captured the nasty tone of many of the messages I received at the time. In it he wrote, "You'll be known as a has-been who rode the wrong horse and became the laughingstock of science for the rest of your life."

These callous words hit their mark. I knew they were untrue, but the media onslaught left me feeling utterly beaten down. *Could this really be the death of my career?* I was tired, and I wasn't sure I could summon the necessary anger to fight back all over again. I'd long understood that anger and outrage were the essential fuel I needed to claw my way back from the brink. But even more than that, I needed allies. Fortunately, it was at this moment that many of my colleagues finally began to speak up.

Neale Monks, paleontologist from the Natural History Museum in London, was the first to emerge, pushing back against a BBC News story written by Paul Rincon about the *Science* paper and coverage of the most

196 THE LAST EXTINCTION

recent LPI/impact conference. It was like "fishmongers confirming the benefits of eating fish," Monk wrote about Rincon's one-sided report, which claimed there was a "confirmed" link between the KPB extinctions and the Chicxulub impact. Monk pointed out how the review downplayed the ongoing decline of species that occurred long before the impact event. Ignoring the decline of dinosaurs well before the mass extinction, he wrote, was like saying the dinosaurs saw it coming, got depressed, and started dying out. Monk's funny and derisive critique lent me some hope that other sane voices might emerge to counteract the monstrous fraud—and they did, in greater numbers than I ever anticipated.

I believe the *Science* review was the tipping point in the Dinosaur Wars, and for an unexpected reason. I was the main target, but the article also casually dismissed the work of many scientists from a variety of disciplines, ranging from paleontology, sedimentology, geochemistry, volcanology, geophysics, and astrophysics. And they rightfully felt insulted. In the past, I had faced the impactor's attacks with little public support because the many colleagues who agreed with me had little stomach for the abuse they might suffer. But their implicit inclusion in this recent attack seemed to have finally nudged them out of the woodwork.

I received many calls from scientists working in diverse fields, particularly from volcanology, geophysics, astrophysics, and paleontology. They were madder than hell and demanded action. In an unexpected turn of events, Schulte's revenge turned the tide in my favor. The formerly silent majority of scientists rallied to reclaim truth from the fake science of the impactors.

Their support changed everything. It lifted me and roused my instinct to fight back. Our first action was writing letters of critique to *Science's* "review" article to set the record straight. Of the many letters that were written, three were eventually published.[36]

The chorus of outraged scientists reached a crescendo at the European Geophysical Union (EGU) conference in Vienna on May 5–8 (2010) where Thierry and I had organized a session on *Deccan Volcanism and Mass Extinctions*. In front of an audience so large that it overflowed into the hallway, Vincent Courtillot lambasted the authors of the *Science* review for "gross incompetence," and called it "the worst fraud ever published, either by stupid ignorance or planned science fraud." Vincent's speech left impactors in the audience gasping and many others smiling with satisfaction. After praising my team's work and criticizing the impactors omissions and false representation in the *Science* review, he veered into his own Deccan volcanism research and eviscerated what was left of the *Science* review article.

To dismiss Deccan volcanism, Peter Schulte had chosen Vincent Courtillot and Frédéric Fluteau's computer model to claim sulfur modeling proved Deccan volcanism insignificant.[37] This claim was both erroneous and a prime example of cherry-picking data. Courtillot made it clear the gas emission of the Chicxulub impact was equal to just one large Deccan eruption of which there were over one hundred. Therefore, regardless of whether the Chicxulub impact hit Yucatan at KPB or 200,000 years before, there was little effect and no long-term consequences on the atmosphere and climate compared with Deccan volcanism.

As I exited the hall still roaring from Courtillot's attack on the forty-one authors, Jay Melosh of Purdue University, a geophysicist known as ferocious pro-impact fighter,[38] caught up with me and complained that I'd set Vincent up to attack the impact camp. They were clearly unhappy to see I was no longer the only one pushing back.

"Jay," I told him, "Nobody sets up Vincent, he is his own man."

"But I've known Vincent for over twenty years! He has always been courteous and would never attack his friends," he shot back.

I must admit I took great pleasure in replying, "Maybe you should have thought of that before signing your name to a paper that attacked his research based on ignorance."

The EGU was not just a bad day for the *Science* review's forty-one authors, it was also bad for Smit, who was invited to give the final public lecture on the evening of the last day. I was already on my way out of Vienna when he spoke, but Thierry Adatte was still there and went to Smit's lecture. Thierry emailed me the results with glee. The public lecture, held in the largest auditorium with 800 seats, was barely attended by fifty and nobody showed up to introduce Smit, which was customary

198 THE LAST EXTINCTION

for such an honor lecture. By the end, the audience was silent, except for two minor questions. Then someone spoke up and accused Smit of a lecture full of factual errors, telling him he would have benefited from attending the morning session on "Volcanism and Mass Extinctions." In fact, I had seen Smit in that audience, but facts never bothered his impact scenarios.

Smit was even more unhappy that Ted Nield, editor of the London Geological Society's *Geoscientist* magazine, had published yet another paper by my team on May 5, 2010, with a front-page photo of the Deccan Traps.

After the Nice showdown, Ted had invited Smit and me for an online "discussion" over the impact controversy which resulted in a back-and-forth that lasted for three months. By the end, Ted became my staunch supporter and always ready to give me advice. He published several of my articles in the *Geoscientist* magazine with frontpage illustrations. He was one of the most funny and witty geologists I've ever met.

In the years following this episode, I often wondered how forty-one scientists had been duped into attaching their names to the *Science* review. Eventually, I got the answer from Rex Dalton, the investigative reporter for *Nature*. He told me that he'd witnessed Schulte make the rounds at the 2009 AGU meeting in San Francisco, signing up co-authors like they were signing a petition. Rex expressed surprise that some "decent folk" caved in to the peer pressure and signed on. It wasn't just Schulte with a sign-up list; I heard from an old friend that there was also a lot of arm-twisting at Berkeley from Walter Alvarez to sign up "co-authors" to give the appearance of consensus science.

Ultimately, the *Science* publication didn't catapult Schulte to new heights in his career. The *Science* paper was a disaster, impactors avoided him as he was no longer useful, and he left academia for groundwater consulting in Switzerland, where he still resides.

Despite the irritating, short-term distraction by the Schulte episode, my professional life continued to surge ahead on many fronts with my students and science collaborators.

Drilling deep cores of the Krishna-Godavari Basin was a tremendous success thanks to ONGC support and their scientists, particularly

Reddy and Jaiprakash. The largest Deccan eruption pulse and longest mega-flows across India left little doubt that continental volcanic eruptions were the major threat to life on Earth. At the same time, our KPB research in northeastern India revealed how marine life was wiped out by extreme warming and acidic and toxic oceans, linked to the fallout from Deccan volcanism. We were back at the cutting edge of KPB extinction research, attracting young and open-minded scientists, who were beginning to view the impact theory as outdated.

Unfortunately, my personal life at this time was not happy. After the fiasco around the Schulte *Science* article forced me to postpone my trip with Rosie and Helen through the Southwest, we were able to reschedule for June 2011. We off-roaded through Grand Canyon, Monument Valley, Canyon de Chelly, and Zion National Park, and celebrated with abandon. Without our husbands holding us in check on such travels, we were wild women, hooting and laughing the whole time. But our adventure was overshadowed by Rosie's evident ill health. Rosie was pale, in pain, and could barely walk, but smiled and laughed through it all.

One afternoon we pulled into the parking lot of a trailhead after a long drive. Helen and I got out and made an exaggerated show of stretching our aching bodies so Rosie wouldn't feel self-conscious taking the time she needed to gingerly unfold herself from the car. Just a few steps from the lot, Rosie sat down on a bench and insisted we go on without her. We didn't want to leave her behind, but it was clear she could go no farther.

"I'll be fine." She waved us away and bent down to tie her shoe. As she did so, the neck of her shirt dipped open revealing a large, swollen lump beneath the skin of her breast.

"Rosie!" I exclaimed in shock.

She had known about the lump and felt ill for more than a year. She suspected it could be cancer but postponed seeing a doctor because she didn't want to miss our US adventure. When we pressed her for further details, she confessed to enormous pain in her breasts and right leg. That evening she let us take a closer look at the lump. It was almost the size of an apple. Nothing was visible on her leg. Rosie downplayed her condition and insisted we proceed with all our fun plans.

200 THE LAST EXTINCTION

"I'll sit in the sun while you two take trips down into the gorges," she said.

Rosie was my dearest sister and closest friend. Just watching her in pain made me sick with worry. Once she returned home to Switzerland, she found out what I had suspected during our trip—she was too far gone. She had Stage 4 breast cancer.

Her first operation took off her breasts. Her doctor promised she still had four good years based on reduced hormone therapy, and she did well. But the pain in her leg got worse, and by the time doctors bothered to take x-rays, bone cancer had eaten away most of her leg. Within a week they operated, cut out as much cancer as they could, and kept pieces of bone, which they screwed onto a two-inch-wide metal plate spanning from heel to knee using sixty-seven screws. It was the first of five leg operations. Through it all, Rosie tried hard to stay positive. I visited her as often as I could, and we still made small trips through the valley of our childhood and laughed ourselves silly with memories of those times.

15

An OUTRAGEOUS IDEA

IN LATE MARCH 2013, 150 INTERNATIONAL SCIENTISTS MET at the Natural History Museum in London (NHML) for a conference on "Volcanism, Impacts and Mass Extinctions: Causes and Effects." The conference had begun in 2010 amidst the professional backlash to the *Science* review article, and its purpose was to set the record straight on the mass extinction debate. In style and substance these gatherings couldn't have been more different than the Snowbird and LPI meetings: No one shouted insults, dismissed evidence, or ostracized those with contrary data and ideas. Unsurprisingly, only a handful of impactors attended. I would have welcomed more of their contribution to the interdisciplinary discussions we engaged in at the conference, where lively debate—conducted in an atmosphere of mutual respect—could turn into learning experiences. There was still so much we needed to know, and the importance of what we were learning was growing more and more apparent.

When Dewey McLean first proposed his theory of Deccan volcanism in the 1970s, he presciently framed it as a warning to humanity about the dangers of human-caused climate change. Now that scientists in the field were seeing how the gradual release of volcanic greenhouse gases had contributed to climate change and mass species loss millions of years ago, this line of research was also underscoring the dangers of climate change to the modern world.

202 THE LAST EXTINCTION

Drawing upon the work of geologists, geophysicists, geochemists, volcanologists, sedimentologists, paleontologists, and astronomers, the conference presented a wide range of research establishing the link between the time frame and eruption patterns at the Deccan traps and global climate change. Robert Spicer and Margaret Collinson described how plant fossil records from the Americas and New Zealand showed extinction patterns consistent with environmental instability over a long period of time. But they had little doubt that the Chicxulub impact caused terrestrial ecological trauma in middle latitudes of North America. Bandana Samant and Dhananjay Mohabey presented the results of studies of sedimentary beds in the Deccan Traps that showed that Deccan volcanism, during its peak period of eruption, played a significant role in the global mass extinction near the K–P boundary.

The conference also presented new research into dinosaur extinctions. It was already known that well before the peak volcanic eruptions thousands of baby dinosaur eggs never hatched because their shells hardened in the toxic environment and they died. This was a harbinger of the coming end of the KPB mass extinction, but David Archibald, one of the foremost dinosaur extinction experts, deepened our understanding of what happened globally by looking at fossils found far from the Deccan Traps, including North America. The dinosaur fossil record of North America is unique with terrestrial vertebrate fossils spanning the last ten million years of the Cretaceous scattered throughout the western interior of the continent. Non-avian dinosaur species decreased nearly 50 percent, from forty-nine to twenty-five species. The decrease began at the Campanian-Maastrichtian boundary (seventy-two million years ago), which was a time of global cooling, when the sea level fell and there was reduced coastal plain environments for non-avian dinosaurs. As time passed, Deccan volcanic eruptions increasingly drove climate warming over the last 300,000 years before the KPB mass extinction. Nevertheless, scientists believed the Chicxulub impact undoubtedly played a major role in the KPB mass extinction.

In 2013, the major question—What really caused the KPB mass extinction?—was still hotly debated. Was it Deccan volcanism? Was it the Chicxulub impact? Or was it both? Currently, Deccan volcanism research was well established in India, but paleontologists were still learning about the major pulsed eruptions that began by 66.3 million years ago. And still today the question of whether it was Deccan volcanism or the Chicxulub impact, or both, persists.

Unlike previous conferences I had attended that had been hosted by impactors, a wide spectrum of conclusions about what caused the fifth extinction was aired at the conference. Though few dogmatic impactors attended, the idea that an impact played a role in the mass extinction was well represented. Some, like me, suspected that volcanism was likely the primary cause of the extinction, but many others proposed a synthesis of the theories. Nevertheless, I welcomed the viewpoints of my fellow scientists and remained curious about any new facts that were brought to the debate. This was how science was meant to be conducted—with serious, evidence-based research presented and debated in a respectful way.

The conference became a blueprint for mass extinction meetings throughout the US, Europe, Asia, China, and India, which are still used to this day. As more young students began to research these topics, the impact theory began to lose its prominence as the sole explanation for the mass extinction. I was pleased to see new arguments and debates emerge after all this time. Even the hard-core impactors began to revise their theories to incorporate volcanism into their asteroid-induced catastrophe scenarios. In April 2014, a team from UC Berkeley, including Mark Richards, Michael Manga, Paul Renne, Walter Alvarez, and others, hurried to the Deccan Traps in India. After briefly studying the site, they argued that the impact (Chicxulub) likely triggered the most immense lava eruptions in India. Furthermore, they explained that the close coincidence between the Deccan Traps and Chicxulub impact had always cast doubt on the theory that the asteroid was the sole cause of the end-Cretaceous mass extinction. This was news to me; several more modifications of this "theory" followed, each reeking of desperation.

My team began to feel we had the wind at our backs. But there was more to do. We still needed to solve one of the most difficult problems: dating the precise age of the pulsed lava eruptions across the KPB mass extinction. This gap in our knowledge meant we could not definitively link climate warming to volcanism, and the mass extinction directly to

Deccan's longest lava flows. Precise Deccan age dating had remained impossible for decades.

Geology fieldwork crew on age dating Deccan Volcanism in 2013: from left to right, Kyle Samperton, Blair Schoene, Mike Eddy, Syed Khadri, Thierry Adatte, and Gerta Keller. On this first field trip to India, I treated my crew to first class hotel and meals to avoid getting sick. It didn't go well. Geologists are not comfortable in grand hotels. You can see their discomfort already upon entering the plush hotel lobby and even more so when they return in the dark after a day of sweaty, dusty, smelly fieldwork hauling large bags of rocks through the lobby. After two days, my crew moved to a small hotel in Pune, no worry about hauling rocks, postponing smelly showers, and dropping dead of exhaustion before dinner. It was a great trip.

Why was it so difficult to date the Deccan Traps? Because of a mysterious absence at the site of zircon crystals. Zircon crystals, which are common in ash layers deposited by volcanic eruptions, are the basis of one of the most common and precise age-dating techniques used by geologists. When they are first formed, these zircon crystals contain uranium, which, over time, decays into lead. Because we know the rate at which this occurs (the uranium ^{238}U decays to lead ^{206}Pb over a half-life of 4.47 billion years) we can reliably date the crystals—and the rock layers they are found in—by measuring the ratio of lead to uranium present in a discovered sample. It is a very accurate way to assess geological age and can be used to date rocks as old as 4.5 billion years. For events occurring around

AN OUTRAGEOUS IDEA 205

the fifth KPB mass extinction that occurred about sixty-six million years ago, it has a margin of error of only ten to 20,000 years.

Three earlier mass extinctions associated with volcanic eruptions had already been dated with this method. So why not the Deccan Traps? Where were the zircons? I was no expert on radiometric dating or on locating zircons, but if you stare long enough at the Deccan Trap mountains in the Western Ghats, your eyes will eventually home in on the red clay layers, known as "red boles," which appear between some lava flows. These were deposited at times when lava eruptions ceased temporarily. I knew from studying the history of geological exploration in the area that these red boles had not yet been examined successfully for zircons. It was a long shot, but I believed we might find zircons there, coming from weathered basalts or from explosive volcanic fallout originating from other parts of the Deccan volcanic province. I was excited by the idea but knew it would be a battle to convince my colleagues to put in the long days and difficult work when nothing like this had been tried before.

I ran the idea by Majda one day. He cupped his chin in his hand and thoughtfully commented: "Liebling, you might be right, but you won't find anyone who'll collaborate on a wild guess, even an educated wild guess."

He was right. But it was a good sign that he thought there was common sense to it. I decided to try this idea on my young colleague Blair Schoene, who operated the most advanced zircon dating lab in the US. Blair had been working on volcanism connected to older mass extinctions but never ventured into Deccan volcanism. As a junior faculty member, he was initially reluctant to act on my wild guess, but the prospect of an all-expenses-paid trip to India and an opportunity to help solve an enduring geological mystery finally proved persuasive.

I arranged the field trip for early December. Our dating team consisted of Blair, his PhD student Kyle Samperton, his postdoc Mike Eddy, and our India guide geochemist Syed Khadri from Amravati University. Rounding out the team was Thierry, my undergrad student Preston Kemeny, and me. Majda accompanied me and visited the Indian Institute of Tropical Meteorology (IITM) in Pune during the time of our fieldwork. We planned afterward to make a trip to the Ajanta and Ellora Buddhist

206 THE LAST EXTINCTION

cave temples cut into the Deccan Traps. To Majda's delight, IITM made his hotel arrangement at the five-star Sheraton Le Meridien in Pune. I decided to treat our team to reservations at the same hotel. We met Majda for long sumptuous dinners with great conversations and laughter for the first two nights.

Geologists are not comfortable in grand hotels, and before long, my crew grew restless. I could read the discomfort on their faces when we entered the plush, sparkling lobby after a long, sweaty, smelly day in the dusty hot outdoors, hauling shopping bags full of rocks. Judging from the looks we received, we must have seemed like a bunch of homeless people ready to camp out in the lobby. Even after we showered and put on clean T-shirts, the team still looked painfully self-conscious among the posh clientele in designer clothing. After two days, my crew announced their preference to move to the small business hotel in Pune at which my team usually stayed. It was a friendly place that welcomed us regardless of how dirty, dusty, and smelly we were. No shower needed before dinner. I stayed with Majda at the Le Meridien and every morning one of our drivers picked me up to meet the rest of my crew.

We chose the Western Ghats of the Deccan Volcanic Province with the highest mountains of accumulated lava eruptions, which Vincent Courtillot's team had studied for over a decade. We followed their footsteps, led by Syed, who had also been their guide. We began our zircon search on the plains near Mumbai, where the coastal volcanic rocks eroded or dipped into the ocean. From there, we made our way up into the mountains where the rock faces loomed ever higher above us.

The Deccan Traps are among the most breathtaking mountains I've seen. Unlike ranges such as the Alps, Atlas, and Andes mountains, which are thrust upward by Earth's shifting plates, the Deccan Traps tower into the sky in layer upon layer of horizontal basalt flows. In the morning and evening light, the layers glow a spectacular crimson red. Standing among them, I felt euphoric. Everyone was excited to begin the fieldwork. We knew that if we succeeded in finding zircons, careers could be made. Blair got a kick out of working on science related to the dinosaur extinction, which thrilled his five-year-old son back home.

AN OUTRAGEOUS IDEA 207

Enthusiastically, the crew started by searching for zircons in the normal way, by looking for ash layers in basalt flows. The mood quickly soured after two days passed with no zircons and only one ash layer. I noticed that no one paid attention to the red boles. We shared an awkward dinner where my entire crew grew testy with me for suggesting such a crazy idea and dragging them to India in the first place. The next day was worse—I could feel a mutiny was brewing. Sitting around a table drinking sodas before noontime, the entire team had dour, dejected expressions, which I captured on camera. They were ready to throw in the towel, give up the planned fieldwork, and take off on a sightseeing tour. I couldn't stand their pessimism any longer and spoke up.

"All you've done so far is look for zircons in basalts and ash layers. Those are very rare and that's not why we came here. If you recall, we came to collect zircons in red boles. I don't care whether you believe my idea about zircons in red boles is crazy. The only way for us to find out is to collect those red boles and process them. Then we'll see just who is crazy."

A long silence followed. Finally, Kyle spoke up: "There's a chance Gerta's right. We haven't taken a real look at red boles yet. Let's start collecting them."

Reluctantly, everyone hauled themselves up from the table, piled back into the two cars, and Syed directed the drivers to return to the bottom of the exposed Deccan Trap basalts. From there we started looking for red boles, which were easy to spot because their color stood out in a red ribbon between the basalt layers, sometimes in a thin red band, and in other places up to two to three meters thick. The thickness of red boles indicated how much time passed without the lava eruptions.

For the rest of the day, nobody complained—at least not in my presence. Whenever we found red boles, we split up into two groups. Blair, Kyle, and Mike collected red boles for zircon analysis. Thierry, Preston, and I collected closely spaced samples of entire red bole layers to analyze the clay content and mineralogy to determine their origin. Each sample taken by my group was a small handful packaged in a small plastic bag and labeled by outcrop name and numbered in sequence. The zircon samples, however, needed to be large, at least two kilograms each, to

208 THE LAST EXTINCTION

increase the chance of finding the rare crystals. Syed, meanwhile, roamed between our teams, helping in sampling and answering questions as to where we were in the Deccan sequence based on Vincent's findings, and locating the precise outcrop positions on Google maps. By late afternoon, morale had improved, and it grew better day by day. By the end of the field trip, our fingernails were stained henna-red from working in the red clay.

When we were finished, Blair's team bagged 160 kilograms of red boles for zircon analyses in eight boxes, each forty-five by forty-five by sixty centimeters in size. Each box held twenty kilograms of sediments packed in ten plastic bags at two kilograms each. The large sample size was necessary to find rare zircons. We shipped the samples back to Princeton via India air mail, which is usually safe and fast but takes a whole day to complete shipping procedure. It begins with trolling the small-town market shops door to door and begging for cardboard boxes the night before, because there's no other way to get them out in the boonies. We hauled the packed eight boxes off to a post office in the nearest city for overseas shipment, where the content was inspected, weighed, and documented for shipping. Then our crew carried the boxes to a nearby sewing shop to be hand-sewn into white cloth for shipping, which took hours. The white clothed boxes were carried back to the post office and the shipping address written in black Sharpie on the boxes. The post official then stamped the hot red seal on each box and plastered it with postage stamps before placing them in a pile ready for shipment. We were lucky: The first two boxes arrived faster than our return plane and the rest arrived during the following two weeks.

Back in Princeton, Kyle immediately began processing the samples. One morning, a few days after our return, I was sitting in Guyot's Great Hall chatting with one of my students, when Kyle ran up, breathless.

"We have zircons," he interrupted, a sly grin on his face, his chest heaving.

Startled, I let it sink in. A wide grin spread over my face as I raised my fist and pumped it up and down victoriously. "Excellent!" I cheered, while Kyle laughed with excitement. My outrageous idea about finding zircons had been right.

AN OUTRAGEOUS IDEA 209

The Princeton dating team walked on clouds with constant happy smiles, but Blair warned everyone that finding zircons was only the first step; it would take weeks to run the analyses and find the age and accuracy of the data. Until then, he kept reminding us, "It's too early to celebrate. And keep it a secret. This can't leak out before the data is in and the paper is submitted."

On April 26, 2014, Blair sent an email to our field team with the subject line TOP SECRET. In it he wrote: "Hi everyone, I just analyzed four zircons from DEC13-30 [date of collection and sample number] from the Jawhar Formation." In geosciences, rock units or Formations (Fm) are given names after nearby towns for easy identification. The Jawhar Formation with its ash layer and red boles was all the way toward Mumbai and just two hundred meters above the base of the exposed lava flows. The email went on, "Two zircons were overlapping at 66.230±0.03 Ma. There are only about fifteen zircons total in this sample, so I'll dissolve all the others and see what we get. Potentially pretty exciting."

Two of the fifteen zircon crystal samples showed an age of 66.23 million years, plus or minus 30,000. For me, "pretty exciting" was an understatement. Already, the results were everything I had dreamed of. But the zircons from the red boles still had to be analyzed. Would they confirm this age? What else might they tell us?

Within a few days, the substance of Blair's "TOP SECRET" message had leaked to Berkeley's impact group. Phone calls from an unknown number began to ring in Blair's office hourly, but no one ever left a message. Eventually, Blair Googled the number and discovered it belonged to Paul Renne, the fervent impact theory supporter at Berkeley who specialized in another radiometric dating method, known as Argon-Argon (Ar-Ar) dating. Out of curiosity, Blair dialed Paul and asked what was up. Paul, it turned out, wanted to talk about the Deccan Traps. Blair admitted that our team had traveled to India hoping to date the Deccan with U-Pb zircon dating but, in a sly bit of misdirection, told Paul we hadn't found any zircons in the basalts. He didn't reveal anything about the zircons we had found in the ash and red bole layers.

Paul Renne told him that the Berkeley team had collected about fifty samples for Ar-Ar dating of the Deccan Traps and were now "plowing

210 THE LAST EXTINCTION

through them" in their own dating lab. Ar-Ar dating is an order of magnitude less precise than U-Pb zircon dating, but Blair was alarmed that the Berkeley team was working speedily toward a similar age dating effort. After the call, he urged his team to double their efforts, proclaiming: "The race is on!"

The remaining eleven zircons dated older at 66.288 million years with an error margin of 27,000 years. This was an exciting confirmation of the Jawhar age close to the onset of the Deccan volcanic eruptions that began global climate warming. The next zircon U-Pb ages came from three red boles and one ash layer yielding ages of 65.661 to 65.535 Ma with the same error margin in the early Paleocene. These discoveries were exhilarating and exciting. We now had some real ages telling us when the climate altering Deccan volcanic eruptions began and how long they lasted in the Western Ghats.

Blair was an associate editor at *Science* and submitted the paper there because it was the top science magazine that could give this discovery the widest circulation. He concluded that the new uranium-lead (U-Pb) zircon age dating revealed intensive Deccan eruptions began approximately 250,000 years before the KPB and pumped out over 1.1 million cubic kilometers of basalt over approximately 750,000 years. These eruptions were consistent with the latest Cretaceous environmental changes and biologic turnover that culminated in the marine and terrestrial mass extinction. And then we sat back and waited for the *Science* paper review. In the meantime, we planned the next field trip to India to collect many more red boles.

During these exciting professional developments, a catastrophe struck my happy personal life. The morning of November 17, 2014, began like any other. I prepared our standard breakfast of steel-cut Irish oatmeal with blueberries, bananas, and almonds when I heard a crash from the bedroom. I ran back and found Majda lying on the floor, trying desperately to get up using his left arm. His right arm hung at his side lame, unmoving. He tried to speak but slurred the words. His eyes, wild and afraid, locked into mine. *Oh my God!* I thought. *Paralyzed, the right side, severe stroke. Stay calm. Save him!*

AN OUTRAGEOUS IDEA 211

The ambulance arrived in five minutes, and I followed it to the Princeton Medical Center. A doctor told me Majda had a severe stroke from a blood clot caused by atrial fibrillation or AFib. "He will survive," he said, "but may be paralyzed. He will have to relearn everything, how to talk, eat, swallow, and walk, if possible, but he may not recover his memory."

I went numb. Majda was just sixty-five years old. He couldn't lose his remarkable, agile mind. His work in mathematics was the one thing he loved and enjoyed most in life, besides me and our time together. If he was to lose his memory, he'd lose everything. I couldn't let this happen to my soulmate. Rather than cry, I decided to bring Majda back. I had once read an article that after a stroke the brain can heal its broken pathways if intervention occurs quickly. The longer one waits, the more the brain is lost for good. I would start working on his memory right away.

The next morning, I brought my computer to the hospital, with all the photographs I'd taken on my travels with Majda. He used to never pay any attention to photos saying, "I keep memories in my brain, I don't need to look at photos." Now I was going to try to use them to stimulate his memory. We got off to a poor start. When I entered his room in the intensive care ward and sat beside his bed, he asked "Who are you?" The next day he recognized me as "G" and asked, "Who is Majda and the man next to G in the photographs?"

He didn't recognize himself or his name. Every day I spent as many hours as he could take with brain training, and it paid off. Within three weeks, he could recall places we had visited in the photos and remember details of the trips. The doctors were astounded by his progress. But his academic memory lagged and frustrated Majda. The doctor told him it would take another six months for that part of the brain to recover, and even then total recovery was very unlikely. For Majda, this was unacceptable. Every day he spent hours trying to read his math papers but failed to understand them. He kept at it, often frustrated, but slowly his math memory came back.

"I'll be back in the game," he told me.

One morning in early April, Majda walked into the kitchen and announced: "G, I've just had a great revelation this morning. I solved

212 THE LAST EXTINCTION

this difficult math problem I had worked on before the stroke. When I got up, the answer flashed in my head. It was so simple and clear! I'm going to write up this paper today."

It became one of his most celebrated contributions to Applied Math. From that day on, he was not only back in the game, he also became a fanatic.

"It's as though a light went on in my mind and ideas tumble out and I solve problems more quickly than ever before."

Majda wrote thirty publications that year, twice as many as ever before, and I asked him to slow down.

He said, "I can't, I may never have such a streak of clarity ever again."

Majda's brain was back, but his right arm had started to shake; he was diagnosed with advanced Parkinson's disease. He ignored it and dove back into science. He had so much more to accomplish and no time to lose. He would live and enjoy his life to the end. We celebrated his stroke recovery with a trip to Switzerland and visited Rosie. Elated to see Majda, Rosie hugged him tightly, then pointed to a candle in front of his photo in an alcove and said, "I've kept a candlelight vigil for you every day since your stroke so you would recover fast."

It was quintessential Rosie, always thinking of others when she needed help the most. Her shriveled right leg gave her terrible pain and she could barely walk. Somehow, she still managed to spoil Majda and me with her incredible home-cooked meals. My memories of that visit are sharper and clearer than others. The tastes and scents of Rosie's kitchen feel alive in my nose and on my tongue, and the sound of our banter rings in my ears. I sensed the three of us were on borrowed time together, so I soaked up every beautiful moment I had, surrounded by the two great loves of my life.

In the meantime, *Science* published Schoene, et al.'s Deccan dating study and a feature story titled "Back from the Dead: The Once-Moribund Idea that Volcanism Helped Kill Off the Dinosaurs Gains New Credibility," by Richard Stone, the *Science* editor reporting from India. The accompanying photo of me with the Princeton Geosciences Dinosaur was captioned, "Gerta Keller's latest findings have pulled her from the fringe back to the frontier."

AN OUTRAGEOUS IDEA 213

That *Science* even acknowledged my role was astounding, and I believed this was solely due to India's *Science* editor.

One thing was sure, Deccan volcanism could no longer be dismissed as inconsequential in the mass extinction debate. Paul Rene acknowledged this much to Richard Stone: "This is an important paper that shows the action happened over a limited time range. The pendulum has certainly swung" toward a role for volcanism.

At this point, most scientists agreed the link between the KPB mass extinction and Deccan volcanism was difficult to deny. But many still doubted that volcanism could have been the primary cause of the fifth mass extinction, even though it was commonly accepted that volcanism was the cause of the other four. The stubborn zombie of the impact theory continued to shuffle along. After all, the common refrain from impact holdouts went, *how could so many be so wrong for so long?*

Within a year of the *Science* publication, our team returned to India in search of more zircons to date the entire Deccan eruption sequence. Our main goal was to identify and date the largest lava eruption pulse that led to the mass extinction we had documented between the longest lava flows in the Krishna-Godavari Basin. The mood of this field trip was very unlike the first when the entire team gloomily labored with little hope of success. Now everyone was hopping with excitement, raring to go as early as possible in the mornings and working until it was too dark to continue. We began with the oldest part of the Deccan Traps and collected every single red bole layer regardless how thick or thin. We searched and collected red boles along the main roads, on small side roads, on isolated hills, in quarries, the parking space of a restaurant, near private homes, and in the Sinhagad Fort. We collected a total of 141 red boles, missing one that was exquisitely incorporated in the carvings of the ancient Karla Buddha Cave Temple near Lonavala. We shipped the samples back to the US in twenty-five boxes of twenty kilograms each. Back in the lab, the red boles resulted in twenty-four good zircon ages, dating the interval up to the KPB, and twelve in the early Paleocene. This was an astounding first-ever feat of high-resolution age dating for an entire sequence of volcanic eruptions spanning 750,000 years.

214 THE LAST EXTINCTION

But what we had was still not enough to comprehensively make our case. For some critical red boles, additional kilos of sediments had to be collected to gather enough evidence. That's where my former PhD student Jahnavi Punekar, associate professor at the Indian Institute of Technology, Bombay (IITB), came to the rescue. Given the coordinates and photo of the outcrop, she collected fifty-four kilograms of red boles. Everything was going great until politics intruded on our scientific endeavors. The newly elected Prime Minister Modi, in an attempt to control graft, shut down the currency for nearly two months, bringing our work to a standstill. The ATMs were out of order and banks issued just twenty-seven dollars per day per person. The post office now only accepted cash. Two days of waiting in infinitely long bank queues permitted withdrawing only 10 percent of the 500 dollars needed to ship our samples to Princeton. Eventually, a courier service accepted an electronic money transfer for payment and the red boles were finally on their way to make history.

Impactors, of course, were unhappy with the more gradual mass extinction scenario described by our evidence, which showed that Deccan volcanism intensified climate, heated it to extremes, and killed off most life. Impactors sketched an alternative scenario, insisting the Chicxulub impact caused the mass extinction due to dust clouds, global cooling, freezing, and darkness with the main source in the North Atlantic. There is some truth there, but only insofar that the North Atlantic continually carried away sediments through the Gulf Stream current, leaving nothing more than remnants of leftovers from past erosions. Ignoring this truth, impactors continued to relentlessly promote the North Atlantic as the true scenario of global cooling causing the mass extinction. For this, there was not a shred of evidence. It's well known that climate cooled about 50,000 years before the KPB mass extinction.

In the North Atlantic, the Gulf Stream continually carried away sediments and dumped it elsewhere. And so, they attributed the global cooling 50,000 years prior to the KPB mass extinction to fit the impact theory. But still impactors dispute the obvious: Climate warmed dramatically during the last 20,000 years prior to the KPB mass extinction. Consequently, the Atlantic Ocean cannot be attributed to the Chicxulub

AN OUTRAGEOUS IDEA 215

impact. The most complete records are well known from Tunisia, Israel, Egypt, and Texas. These are areas of undisputed complete and nearly complete sedimentary records that are not influenced by the Atlantic Gulf Stream erosion. And still impactors believe the incomplete North Atlantic record as the true cold event of the KPB mass extinction and consistently ignore the true data of the extreme climate warming.

The Gulf Stream erosion in the North Atlantic is missing the pre-KPB age. But in the Mediterranean, Israel, Egypt, Tunisia, northeastern Mexico, and Texas, there is no Gulf Stream erosion and sedimentation was normal.

Despite this soft-peddled approach, it was clear to those who understood the data that the impact theory had lost more scientific ground. That didn't deter editors at *Science* from attempting to put a finger on the scale. I could understand their reluctance to let go of the theory they had championed for so long but was again surprised at the shenanigans they were willing to employ to influence a scientific debate. Under normal circumstances, our findings would be published, after which other scientists could interrogate our data and publish their own findings. Instead, they delayed publication of Schoene's paper for seven months to allow Courtney Sprain and Paul Renne to finish their Ar-Ar dating and write their opposition paper. When their paper finally came out in *Science* 2019, they argued that 75 percent of Deccan eruptions occurred *after* the KPB mass extinction, and the 25 percent of the eruptions that preceded it were not related to global climate. Hence, they argued, the Chicxulub impact must have triggered the accelerated Deccan volcanism.

These conclusions were so wrong it was mind-blowing, and, amusingly, their argument rested on the imprecision of their data. The margin of error for Ar-Ar dating is at least one magnitude larger than U-Pb dating, or about plus or minus 200,000 years. By taking the lowest age from this margin they could date most of the eruptions after the mass extinction. Of course, it was impossible to know exactly how they arrived at their figures because they did not show their data. To add insult to injury, the *Science* editor would not allow Schoene to respond to these erroneous charges.

216 THE LAST EXTINCTION

Science published both papers in the same issue along with an opinion piece advocating for the impact side. Unsurprisingly, science journalists who were unable to parse the technical arguments went along with *Science*'s editorial opinion. Chicxulub was resurrected once more while Schoene's high-resolution Deccan study was thrashed. *Science* had again sandbagged a scientist who dared follow the facts and question impact theory, just as they had sandbagged me with their farcical "review" in 2010. It was a depressing setback for Blair, as he experienced a dose of what I had been through for decades. He couldn't rally his anger to fight back immediately, but he did eventually do a rigorous analysis of Renne and Sprain's Ar-Ar data and compare it to his U-Pb zircon data. Predictably, it showed the absurdity of the impactors' case. By then it was too late. The impact zombie science was reanimated once again. The Dinosaur War raged on.

16

The TRUTH
About CHICXULUB

LATELY, THERE ARE MANY REASONS TO DESPAIR ABOUT THE power of facts and rational thinking to change minds these days. Our complex world is awash in fake news. People are skeptical of authority. And many have lost faith in their own ability to assess truth.[39] The idea that truth doesn't really matter also seems more widespread than ever.

In the light of these disturbing trends, I take great satisfaction in knowing that the evidence others and I have gathered over the past few decades about what really caused the fifth extinction has changed minds. Not long ago, there was widespread belief that the death of the dinosaurs, along with so many other species, was solely the work of a rogue asteroid. Today, scientists who were once skeptical of other explanations for the fifth extinction, even those who once vociferously advocated for the impact theory, concede that Deccan volcanism played a key role.

It is my hope that eventually the general public will also see the subject clearly and realize that the famous asteroid impact they have heard so much about happened long before the mass extinction, and that it was never powerful enough to have caused mass species loss. It takes time to correct a myth that has been spread for nearly forty-five years, especially one as entertaining as the impact theory.

But scientific theories must be based on more than just being a good story. They need to be backed up by facts. We can never definitively state

218 THE LAST EXTINCTION

that a scientific theory is true, only that it supports all the evidence that has been gathered. But what we can be sure of is if evidence contradicts a theory, then that theory must be adjusted or abandoned. It is understandably hard for scientists who have spent their careers in the service of a discredited theory to abandon it. But scientists should not be acting in service of theories, we should be in service to the truth.

In this book, I have presented many strands of evidence that show that the impact theory is untenable. These include several different approaches to dating the moment of the Chicxulub impact, each of which have consistently revealed that the asteroid strike long preceded all evidence of mass species death.

The book has also presented a counter-explanation for the fifth extinction—Deccan volcanism. The abundant evidence for this theory not only lines up with the fossil evidence on species death, but it also describes a more gradual mechanism of global climate change that is consistent with other mass extinctions in the planet's history.

Compiling support for a scientific theory is an ongoing process, and one that continued as I worked on this book. As I was writing, my colleagues and I were studying the latest piece of the puzzle that shows us what happened sixty-six million years ago. Before I present this recent piece of evidence, let me briefly recap the scientific case against the impact theory.

As you'll recall, a series of discoveries in Mexico in 1999, 2000, and 2003 demonstrated that the Chicxulub impact occurred well before the KPB mass extinction. Yes, a large meteorite approximately 175 kilometers in diameter crashed into Yucatan, Mexico. Yes, the impact was a major catastrophe. But based on the fossil evidence in the Western Hemisphere at the time when it hit the planet, we can see that its effects were not substantial or lasting. On the scale of large impacts, it didn't cause significant species extinctions or long-term climate and environmental effects anywhere from North America to Mexico and Central America. Life continued despite the impact's short-term interruption, and rapidly normalized. The impact was a mere blip in the violent geological history of our planet, the true extent of which might never have been understood, had evidence of the event not been discovered in the two-meter-thick

THE TRUTH ABOUT CHICXULUB 219

impact spherule layer that remained undisturbed in quiet deep waters at El Peñón in northeastern Mexico.

Discovered in 2000 by my students Richard O. Lease, Steven A. Anders, and Payan Ole-MoiYoi, this spherule layer is the Chicxulub impact melt rock that rained from the sky and settled to the seafloor in less than one hour after the impact. Using paleontological age dating of the planktic foram species *Plummerita hantkeninoides*, the spherule fallout was about 200,000 years pre-KPB.

We weren't the only ones to make this finding. Two Swiss students, Mark Affolter and Lionel Schilli, also studied a site in the Loma Cerca locality, about twenty-five kilometers from El Peñón. They confirmed a pristine fifty-centimeter-thick impact spherule fallout that also dated 200,000 pre-KPB.

The critical age control helped us also date the spherules prior to the Chicxulub impact and volcanic event. This was the evolution of the planktic foram *Plummerita hantkeninoides*. This evidence came from the Chicxulub drilling expedition known as Yax-1 (Yaxcopoil-1), which was performed from 2001 to 2002 in a location about forty kilometers southwest of Merida, Mexico. The Yax-1 samples showed that sediment on the seafloor was colonized by invertebrate species, including the foram *Plummerita hantkeninoides*. This foram species perished in the KPB mass extinction. If a meteor strike caused the KPB, then how could this species of foram be present in the impact crater? Despite this age evidence, impactors believed the presence of *Plummerita hantkeninoides* in the fifty-centimeter core was due to the Chicxulub impact tsunami, which dredged up older fossils and moved them to the KPB layer.

Thus, we had two outcrops in distant regions, studied by separate teams, where each confirmed that the Chicxulub impact predated the KPB mass extinction by about 200,000 years. The samples had been taken and analyzed according to well-established geological practices. The conclusions were based on the evidence, straightforward and conventional. Nevertheless, impactors stubbornly resisted what these multiple lines of evidence were clearly saying. They needed to explain away the conclusions. And so they resorted to a theory that was as novel as it was convenient: The Chicxulub asteroid caused the KPB mass extinction,

220 THE LAST EXTINCTION

and the evidence that contradicted that belief was due to a "KPB impact tsunami" that mixed up the sediment layers.

Now consider these two competing theories. One says that the reason these two separate spherule deposits, as well as the crater, all shared the same pre-KPB date is because that's when the Chicxulub asteroid hit.

The other theory asks you to believe that the dating of the crater and both spherule deposits were wrong, and that these errors were due to a massive tsunami that mixed up the earth in a way never observed. Not only that, but this tsunami embedded two different spherule deposits in sediment layers that happened to share the exact same date, a date which coincided with the incorrect Chicxulub crater date, in locations twenty-five kilometers from each other.

As recalled earlier, at the 2003 EGU-AGU Conference in Nice, France, my team disproved the Chicxulub crater age along with evidence of six other sources in the crater core Yax-1. The evidence we presented was devastating to impactors and the famous conference was rarely mentioned again. But impactors (Smit, et al.) revived the same false story and kept repeating it to this day.

All this evidence should have forced impact scientists to abandon their theory that the Chicxulub impact caused the KPB mass extinction. But the impact theory persisted, thanks to its exciting, apocalyptic narrative: an asteroid crashing into Earth, causing global wildfires, freezing, and darkness like the nuclear winter, followed by the mass extinction of the dinosaurs along with 75 percent other life on Earth! It's an exciting story, rich with fantastic images that appeal to both children and adults. But it was not based on science. It never captured the truth of what really happened to the dinosaurs and many other life forms.

Which is not to say the impactors were acting in bad faith. While they were not open to any of the mounting research that showed the impact theory was wrong, they did believe their theory was correct and were willing to continue looking for evidence to support it. And so, in 2004, a small group of impactors lobbied to redrill the Chicxulub cores. Would new samples finally provide the hard evidence they needed to support their scientifically challenged theory? It would take until 2017, when their application for a new drilling operation was approved, for us to find out.

THE TRUTH ABOUT CHICXULUB 221

REVEALING THE TRUTH ABOUT MEXICO . . . AGAIN

In 2018, the new coring Expedition 364, "Chicxulub Drilling of the KPB," was launched. The redrilling was done by the European Ocean Drilling Expedition (IODP), which was funded mainly by Germany, but led by US and British science teams. In its secrecy, this expedition was unlike any that had come before. Core samples were kept under lock and key by COPI Sean Gulick from the University of Texas at Austin. He controlled who could study the core samples and what studies were permissible from EXP-364, which nobody could access but Gulick and his cohorts. By restricting who could test the samples, and what kind of research was permitted, it had the hallmark of past impact-controlled studies, which excluded many scientists who didn't already agree with their theory.

Then I got a lucky break. In the fall of 2018, an old friend offered me core samples from EXP-364. I was thrilled. "How did you get these samples?" I asked. Nobody has access to these core samples. My friend told me he sampled core EXP-364 at the German archive in Potsdam on the first day the samples became available for study. He collected the most important core samples from the limestone above the impact crater to the black clay above the KPB followed by Paleocene limestones. The most critical part of the impact crater is below the KPB and could make or break the answer to *What really happened to Chicxulub?*

My friend was very fortunate to have collected EXP-364 samples. Already on the next day, the German core EXP-364 closed for unexplained reasons. The core was transferred to the University of Texas at Austin by CO-PI Sean Gulick and sequestered under lock and key. From there on, only Gulick had sample access and a small number of his collaborators to study the cores. Gulick and his colleagues turned out dozens of papers claiming Chicxulub was the KPB impact crater but rarely provided testable results.

And now I had access to EXP-364 core samples. What an unexpected great deal for my lab!

When I first received my friend's EXP-364 samples I saw nothing unusual—and why should I have expected different? EXP-364 and Yax-1 are two virtually identical cores drilled on different sides of the same

222 THE LAST EXTINCTION

crater. We saw the same sediment deposition at approximately fifty centimeters length as well as the same laminated carbonate-rich clays. The intervals of green clay, known as glauconite, looked exactly as one would expect after slow accumulation over tens of thousands of years in quiet waters. In both sample intervals, one can clearly see that animals left their tracks in the sediments as they scavenged for food, creating traces like *Planolites* and *Chondrites* and planktic forams. There was no sign of tsunami deposition. But impactors again claimed this was evidence of the impact tsunami at KPB age. There was no shred of tsunami evidence in EXP-364 or the originally cored Yax-1.

Interestingly, neither the fifty-centimeter core EXP-364, nor the identical fifty-centimeter core Yax-1, drilled at the same Chicxulub location, revealed any evidence of impact spherules, or unusual core sedimentation, or core age. The reason for this mysterious absence of impact spherules in the cores Yax-1 and EXP-364 was unknown.

Fortunately, there remained one test that could draw out some information from the tiny EXP-364 samples I had received. We could analyze the mercury (Hg) in the Chicxulub core and test it against the known massive volcanic eruptions that occurred around that time. This could help date the samples by revealing the times of mercury fallout.

When we did this, we found mercury in the Chicxulub EXP-364 core that could be linked to two major eruptions previously identified as Hg *Extreme Events* EE9 and EE8. We knew these happened 73,000 years and 105,000 years before the KPB mass extinction, respectively. This suggested that long after the Chicxulub crater was created by an asteroid about 200,000 years ago, these volcanic eruptions ejected major mercury fallout that settled in the crater core. There was no way this layer of mercury could have predated the impact. We had, yet again, another piece of evidence showing that the EXP-364 and the Chicxulub impact predated the KPB mass extinction by about 200,000 years.

A scenario in which major upheavals caused by Deccan volcanism resulted in major climate warming was no surprise to us. There is no comparable evidence of species extinctions caused by the Chicxulub asteroid impact EXP-364 about 200,000 years pre-KPB. All foram species survived, and life continued normally.

THE TRUTH ABOUT CHICXULUB   223

What changed was catastrophic climate warming, beginning during the last 20,000 years and accelerating over the last 10,000 years to the KPB mass extinction, which had become unstoppable in India and worldwide. Seventy-five percent of life rapidly perished by hyperthermal warming and toxic environments, ending the mass extinction. By the end, just one planktic foram survivor remained—*Guembelitria cretacea*, the smallest, primitive species and survivor through the ages.

What happened to Chicxulub 200,000 years pre-KPB is well known but hidden and disguised to keep the impact story alive—and it has worked to this day. But truth has a way of revealing itself despite impactors' strongest intentions.

During the Covid pandemic, my students Udit Basu, Casey Conrad, postdoc Paula Mateo, and I decided to build upon the mercury findings at Chicxulub to see what other evidence we could find throughout Mexico. Over two years we analyzed well over 3,000 samples for mercury fallout caused by Deccan volcanism. No such analyses had ever been done before in Mexico because impactors only believe in the Chicxulub impact tsunami story and often prevented publications of our work.

In this study, we found evidence from various mercury Extreme Events (EE) in samples from Mexico, which confirmed the theory that Deccan volcanoes launched into the high atmosphere and spread around the globe. This theory was previously confirmed based on Deccan volcanic eruptions in India followed by the global distribution of mercury fallout over one to one and a half years in Tunisia, Israel, and Egypt (Keller, et al., 2020). We built the new data upon the next step in the global distribution of mercury fallout in Mexico.

There are various age-dating methods to assess Deccan geochronology. The oldest and most simple method is based on paleontology: the age of species evolution and extinctions over time. The most famous example is the evolution of *Plummerita hantkeninoides* at 220,000 years pre-KPB and its mass extinction at the KPB. The most complex, time-consuming dating method is based on high-resolution U-Pb zircon dating with reported uncertainties up to the 95 percent Confidence Interval (C.I.). This is the most successful age-dating method developed by my colleague Blair Schoene and his team based on Deccan volcanic eruptions in India. The

224 THE LAST EXTINCTION

next age-dating method is based on mercury events, EEs, which match up with previously established timelines from Deccan volcanism in India and the global mercury fallout worldwide.

There is also evidence of local climate change resulting from species loss, which suggests that the ongoing climate change from the Deccan eruptions was causing extinctions. However, species loss was gradual over a long period of time spanning about 250,000 years. Only during the last 20,000 years pre-KPB did climate rapidly turn to mass species loss causing the mass extinction of 75 percent of species loss.

All the evidence we found agreed with established geological interpretations of sediments and conformed with our theories that intermittent volcanic activity on the other side of the world was producing spikes of extreme climate change in North America over a time frame spanning several hundreds of thousands of years.

Impactors, as you'll recall, had a novel theory for these types of deposits. They said the layers that didn't conform with their timeline had been mixed up due to an "impact tsunami" from the Chicxulub impact. But geologists know what tsunami deposition looks like. It is always chaotic and occurs over hours to a day or two. In contrast, sediments at El Mimbral deposited over a long time of mixed impact spherules alternating with limestone followed by thick sandstone. Over time, sandstones reduced, and normal sediments resumed along with diverse fossils. The end of deposition was a thin, black clay which followed the next cycle of normal sediments. There never was an "impact tsunami" deposit anywhere in northeastern Mexico. This misconception or lie was perpetuated by impactors (Jan Smit) ever since 1992, despite all evidence to the contrary (Keller, et al., 1993) and repeated endlessly to this day. Truth has been eliminated in favor of belief in the impact theory.

We have analyzed dozens of KPB sections in northeastern Mexico, including Rancho Canales, El Mimbral, El Peñón, La Lajilla, El Mulato, La Parida, and the Chicxulub impact crater core Yax-1 and EXP-364. All of them, except the Chicxulub impact cores, have abundant reworked impact spherules from older deposits. A limestone frequently separates the reworked spherules. There are no undisturbed impact spherules in any of these deposits dated near the end of the KPB extinction.

Nevertheless, impactors claimed the reworked spherules are from an "impact tsunami," but there never was any evidence of impact or tsunami deposition. Sediments consisted of normal marl and sandstones.

And still today, the Chicxulub impact reigns supreme even though the data is incorrect, as we demonstrated all over Mexico based on multiple evidence over many years. But impactors prevailed in their belief to this day. Nobody dared to question the impact theory and so they remained on the safe side. Who will have the courage to speak out for truth? The Chicxulub impact predates the KPB by 200,000 years.

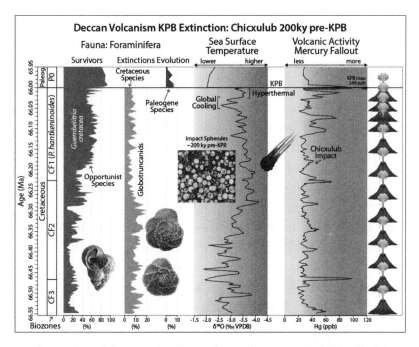

Volcanic activity includes mercury from Deccan volcanism; the mercury is distributed worldwide by winds over one to one and a half years and then settles on land and sea. We can measure and date the mercury fallout based on quantity and frequency of eruptions. This data reveals the major mercury peaks during the past 66.57 to 66.00 million years. Most outstanding are the extreme volcanic eruption peaks that mark the hyperthermal warming and end with a series of nearly constant eruptions across the KPB. This is the mass extinction and its rapid end. There is no evidence that the Chicxulub impact is KPB age, as commonly believed. A multitude of evidence reveals the impact predates the KPB by 200,000 years based on age dating of Plummerita hantkeninoides. More recently, Mercury dating of the eroded Chicxulub structure revealed 90,000 years missing.

226 THE LAST EXTINCTION

These days, there is ample evidence that reveals what happened when the world changed sixty-six million years ago during the fifth mass extinction. And what happened earlier at 66.2 million years ago when a large meteorite crashed into Mexico. And then what happened earlier at 66.3 million years ago when the climate cooled. These are the stories of decades of discovery, showing how the evidence revealed itself step by step as we analyzed the data—and how, eventually, the truth could no longer be denied. But before we dive deeply into this amazing evidence, let me briefly recall some of the impact stories.

At the beginning of the Dinosaur Wars in 1980, it was Luis Alvarez, a Nobel laureate, and his son Walter who claimed that an asteroid or meteorite crashed into Earth and caused the dinosaur (along with other species') mass extinction. It was a good story but lacked critical evidence. In 1991–92, the discovery of impact glass spherules in Haiti, El Mimbral, and El Peñón was an exciting time to research the age, origin, and environmental effects of impact glass caused by the meteorite. Already at that time, impactor Jan Smit quickly added mega-tsunami waves as cause for the Chicxulub impact. While this made a perfect story, there was never any impact tsunami evidence in Mexico, making her theory completely unfounded.

When Deccan volcanic eruptions peaked sixty-six million years ago, hot lava raced through 1,000 kilometers into the Bay of Bengal. Just one planktic foram species, Guembelitria cretacea, and its many offsprings survived and continue to thrive today. But impactors denied the evidence; for them, it could only have been the Chicxulub impact.

While global climate warming is held as a truth across the KPB in northeastern Mexico, Tunisia, Egypt, Israel, Europe, the US, India, and elsewhere, in recent years, impactors have denied it, claiming it was global cooling of the North Atlantic Gulf Stream instead. It is hard to fathom how anyone could take this theory seriously. The North Atlantic Gulf Stream has carried and eroded sediments ever since the North Atlantic opened. Why were impactors not aware of this largest current? Was this a joke? Or was it ignorance?

Eventually, this claim was realized to be so outlandish, that even impactors let it die.

THE TRUTH ABOUT CHICXULUB 227

ANALYZING THE SCIENCE RECORD: PART 1

It is not easy to understand the wiggly lines of science, especially if you are not familiar with the field. In this chapter, my friend and colleague Paula Mateo and I have made great efforts to illustrate and simplify the science of the Dinosaur Wars—the decades of fighting for truth denied by impact theorists. This is not to say that impact theorists always lied. On the contrary, I think they wholeheartedly believed that their theories were true, no matter what came next.

We start our science discoveries with the basics: the wiggly lines of science you see in the four figures. Each figure shows a set of very different but interconnected data that independently proves itself. By the end, you will have learned a great deal of science as the stories unfold. Each figure begins from the base (oldest) to the top (youngest), and along this vertical line are ages in millions of years (Ma), 66.55 to 66.00 (KPB), and the Paleogene 65.95 Ma. Next are biozones from bottom to top: CF3, CF2, CF1, followed by the Cretaceous-Paleogene Boundary Mass Extinction.

We begin with the lower half of the species' record zones, CF3 and CF2, which span from 66.55 to 66.22 Ma and show two species groups: survivors and extinctions. At this early stage, most species survived. At 66.22 Ma, there is a major increase in the abundance of the smallest species: Guembelitria cretacea, in biozone CF2 (column 1). This small, simple species survived through the ages and is still living today. In contrast, the large specialized Globotruncanids (column 2) decreased in abundance 66.22 Ma and never recovered. The cause was climate warming and increased salinity, oxygen, and nutrients. Ornate species that once thrived became rare. The main cause was climate warming (column 3, warming right, cooling left).

The older climate records from 66.55 to 66.35 Ma show extreme climate variations with high peaks but uncertain origins. Analysis from mercury fallout (column 4) shows six Deccan volcanic eruptions. Hence, we know that Deccan eruptions occurred regularly.

We conclude there are no mind-boggling discoveries. Life changed gradually; some species died, others decreased in abundance. The climate

228 THE LAST EXTINCTION

changed from cool to warming. Deccan volcanic eruptions and Mercury fallout left no record that life could change significantly.

SCIENCE: PART 2 – THE CLOCK IS TICKING

After investigating the earlier science results from 66.55–66.23 Ma, we reviewed what happened near end-Cretaceous. Only one new species had evolved: Plummerita hantkeninoides, named for biozone CF1 and dated 220,000 years pre-KPB mass extinction. What other hidden stories could we find?

Then, the most unbelievably exciting event occurred, just 200,000 years prior to the KPB mass extinction. It was an event nobody could ever forget—not the three students digging the trench that hot afternoon in the year 2000 at El Peñón in northeastern Mexico; not my dozen students, digging other trenches nearby and running to see the exciting discovery; and especially not me.

That day, we discovered a two-meter thick impact melt rock welded at the base, followed by glass rafts settling to the sea floor. As the spherules cooled, smaller glass spherules followed so that the cooler impact glass spherules were perfectly rounded and ended up on top (Fig. 5). This was the most exciting field trip in search of truth.

As we dive back into chapter 16, zone CF1, Guembelitria cretacea remains abundant and the sole survivor. The large, specialized Cretaceous species are on their way to extinction, but there is still more data from the impact spherules dated to 200,000 years pre-KPB shown by glass spherules and a short climate cooling (column 3). This cooling occurred because of the Chicxulub impact that we now know occurred at 200,000 years pre-KPB. But this impact that crashed into Yucatan left no significant effects nor species extinctions. Moreover, the Chicxulub impact eroded up to one hundred meters of sediments, as is evident from mercury data. At the same time, we discovered the Chicxulub impact crashed into Yucatan and left evidence from Mercury fallout (column 4). We now have the same data from multiple sources that the Chicxulub impact predated the KPB by 200,000 years.

THE TRUTH ABOUT CHICXULUB 229

There is still more evidence from relatively short global cooling just prior to the onset of massive Deccan volcanic eruptions. The cooling correlates with near absence of Deccan eruptions, which indicates a lull in the eruption sequence. Some impactors claimed that global cooling occurred throughout and was caused by the North Atlantic Gulf Stream, but we know that the Gulf Stream erosion is ubiquitous and erodes sediments consistently and continuously. Gulf Stream erosion and climate change provide no support for impactors' beliefs.

Near the end of the KPB mass extinction, global climate warming turned into hyperthermal warming as Deccan volcanic eruptions flowed over 1,000 kilometers into the Bay of Bengal. Several additional long lava flows followed, and the mass extinction ended. The evidence is clearly seen in the Deccan eruptions and the hyperthermal warming. When the mass extinction ended, so did all but one foram species: Guembelitria cretacea, a survivor still today.

17

EXPLAINING *the* "IRIDIUM ANOMALY"

(CONFESSION OF A FORMER IMPACTOR)

WHEN WE ARE TAUGHT THE HISTORY OF SCIENCE IN SCHOOL, we are encouraged to see each discovery as part of a series of connected breakthroughs, each taking humanity one step closer to the truth of how the natural world works. But our foreshortened contemporary perspective on the finer details of intellectual history can obscure the often-windy path scientists have traveled as they advance human knowledge—and the many wrong turns they have taken, even in the face of compelling new evidence. Today, we can look at forms of different organisms, the way different species interact, and the heritability of traits, and clearly see the overwhelming evidence for evolutionary theory. We can look at a map of the world and see how the continents must have once fit together. These ideas—that living things evolve and that continents move—are obvious to us today, but it took decades before the theories that explained them were accepted by many of the most respected and highly trained members of the scientific establishment.

The truth is, it can be very hard to change a person's mind once they've made it up. And this is especially true if that individual is a scientist. After all, scientists spend years acquiring specialized knowledge, and their entire job is formulating hypotheses and theories, and rigorously

232 THE LAST EXTINCTION

testing them. So it's not a surprise that scientists can be very stubborn about their conclusions—each of us are convinced that we are guided by reason and that our beliefs are informed by our careful evaluation of multiple lines of evidence, guided by our experience and expertise.

Of course, scientists are wrong all the time, and two scientists with the same training can look at the same facts and come up with starkly different conclusions. What I've learned is that if you are interested in the truth, you must be skeptical of your own beliefs, and willing to constantly test them. At the same time, you should always be open to new theories, so long as they are sound and supported by good evidence. What a scientist shouldn't ever do is become so enamored of a theory that they cease to question it. Nor should scientists shut themselves off from contrary evidence, or worse, suppress that evidence so that others can't access it. When I look back at the Dinosaur Wars, I experience a wide range of emotions, but few are as strong as my dismay over how so many scientists, for so long, unquestioningly adopted a theory that had so little evidentiary support. Fortunately, scientists are finally beginning to change their minds about the impact theory as more and more evidence points to facts that Deccan volcanism was the cause of the KPB mass extinction. And Chicxulub predated this impact by 200,000 years prior to the KPB.

The best scientists I have known strive for the humility, curiosity, and rigor that are essential for truth-seeking. And few embody these virtues more than my friend and colleague Munir Humayun. Though we didn't always agree, he has been one of my favorite collaborators and most hopeful examples of how scientific evidence still has the power to change minds.

Munir is a cosmochemist, which means he is an expert in the chemistry of matter in space, as well as how it forms and the different ways it comes to Earth, including comets and asteroids. When we first met in 2017, Munir and I were on opposite sides of the Dinosaur Wars. But that didn't mean we couldn't get along. Munir is great company, and one of the few scientists with whom I could have serious and fun discussions about the causes of the fifth extinction, and particularly the origin of iridium (Ir) in the KPB clay layer.

EXPLAINING THE "IRIDIUM ANOMALY" 233

This "Ir anomaly" is one of the key pillars of the impact theory, and the very first piece of evidence cited to support it, dating all the way back to 1980. That was when the Nobel Prize–winning physicist Luis Alvarez and his geologist son Walter collected samples across the Cretaceous–Paleogene boundary (KPB) in Gubbio, Italy, that contained a distinct dark gray clay layer with a three-to-four-millimeter thin red clay near the base. Lab analysis revealed anomalously high concentrations of iridium in the red clay tailing upward. Alvarez believed the Ir anomaly was the telltale signature of a large asteroid impact, which created a dust cloud that prevented photosynthesis and caused the mass extinction. The proof, he claimed, was the chemical composition of the KPB clay believed to come from fallout of the impact dust cloud.

At the time, I was surprised that this theory, that the Gubbio iridium was extraterrestrial and spread by an impact, was never adequately tested, and that the few studies that were performed seemed to contradict the impact theory. KPB Ir analyses were published from locations across the globe and most frequently revealed small concentrations of iridium—less than 0.5 to 1 ppb (parts per billion)—commonly in the absence of the red layer. Impactors made much of a half dozen localities that revealed high Ir concentrations greater than 8 ppb. But geochemists pointed out that these higher concentrations were always in a thin iron-rich layer. The iridium, they suggested, likely filtered down through higher layers and settled in the red layer, which was impervious to further leaking into lower sediments. A few chemists with knowledge of planetary chemistry (cosmochemists) tested the elemental platinum and siderophile (or "iron-loving" elements) concentrations among the iridium and found them incompatible with an asteroid impact. Nevertheless, in spite of these relatively low concentrations and other conflicting evidence, the mere fact that Ir was present at all was claimed as the impact's proof. And just like that, the idea that the iridium layer was of extraterrestrial origin was widely accepted as fact. Amazingly, no one ever did a deeper analysis of this evidence in the many years that followed.

Some of us remained skeptical. Through the decades of the Dinosaur Wars, I dreamed about reinvestigating the origin of the Ir anomaly at the KPB mass extinction. But I knew I lacked the chemical expertise to

234 THE LAST EXTINCTION

tackle this difficult problem. I had become increasingly doubtful of Ir's impact origin and was leaning toward a volcanic origin. In the 1980s, there had been several studies by volcanologists and geochemists showing that Ir was common in volcanic eruptions. One measured over 500 ppb concentrations of iridium in ash layers from volcanic plume eruptions, including on the Hawaii and Reunion islands, with the latter the successor eruptions to the Deccan Traps. That is sixty-three times the amount of iridium that the Alvarez's found. These eruptions demonstrated that iridium from deep in Earth's mantle could surface from deep-seated volcanism and contribute to the Ir anomaly in surface crustal rocks. Thus, Ir and other elements from Deccan volcanic eruptions could be a likely source for the Ir at the KPB clay. As the years passed, and the Dinosaur Wars carried on, I held on to this hope of applying new techniques to get to the bottom of the iridium anomaly question. But by 2017, I had almost given up finding an expert with knowledge in planetary chemistry to finally test the Ir's origin in the KPB red clay.

Then came my lucky day. On February 24, 2017, Majda and I were invited to give lectures at Florida State University, Tallahassee. My talk was about Deccan volcanism, climate change, and mercury—a provocative mix of topics for any impactors in the audience. So I was surprised to be so warmly welcomed by Munir Humayun. He had great charisma, a friendly smile, was very knowledgeable, and seemed open-minded. Remarkably, this was the first time I had been questioned by an impactor purely out of curiosity and an interest in the science, rather than with hostility designed to undermine the evidence I presented. When, after my talk, Munir introduced himself as a cosmochemist, my eyes lit up. Could this be the one I had searched for decades to solve the Ir problem?

We had a long discussion about the origin of iridium, and I expressed my hope that with his curiosity and knowledge I might persuade him to finally take a closer look at this under-scrutinized evidence for the impact theory. Munir was interested but also hesitant. I could see he was reluctant to get into this research, expressing time constraints, the need for samples, developing methods for investigations, finding a good student, and much more. I urged him to think about it and let him know that my archive of Ir-rich KPB samples was available. I didn't think much over the

EXPLAINING THE "IRIDIUM ANOMALY" 235

next year, assuming Munir understandably did not share my urgency or desire to question one of the fundamental pillars of a widely accepted theory. Then, in February 2018, Munir informed me he had decided to tackle the Ir problem. I was elated. From my large KPB archive, it was easy to send him samples from the best localities at El Kef and Elles in Tunisia, which contain high Ir concentrations. At this point, all I could do was stand back and wait to see what cosmochemistry could reveal about the evidence.

It took a year, but during my next lecture at FSU, Munir and his graduate student Steffanie Sillitoe-Kukas shared with me the first results of their analysis. Their preliminary conclusions pointed toward a non-impact Ir origin.

I was ecstatically happy to have my instincts validated—at least in this early stage. But I was also excited that this enduring mystery was soon going to be solved. Finally, there would be some clarity about Ir's origin. Munir cautioned there were many other tests to be done that could still turn out in impact's favor. Then came the Covid pandemic shutdowns until late spring of 2021. By October that year, Munir and Steffanie presented their findings at the first open in-person GSA conference in Portland, Oregon: The Ir anomaly, they emphatically stated, was not of impact origin. Munir emailed me a summary of his findings, cheekily titled: "Confessions of a Former Impactor."

Munir's title brought a big smile to my face. I never questioned his impact convictions or tried to persuade him otherwise. He did the research and convinced himself that the Ir anomaly could not be from an impact. Like any excellent scientist, he searched for the truth, and when it stared him in the face, he acknowledged the inevitable: the belief that the impact theory had been wrong all along.

Munir Humayun's "confessions" detailed the complex chemical investigations that led to this conclusion, together with his graduate student Steffanie Sillitoe-Kukas. Like any complex science, chemistry has its own language that can be confounding to anyone not trained in chemistry. Fortunately, Munir was able to tell the story of how he and Steffanie solved the last key problem of the impact theory beautifully and simply, step-by-step, as summarized here:

236 THE LAST EXTINCTION

One important prediction of the asteroid impact theory is that the impactor debris is globally distributed by the energy of the impact. Physical modeling of the aftermath of the impact showed that most of the debris was energetic enough to be driven high above the Earth's atmosphere, but still grasped by Earth's gravity that eventually pulls it back to the Earth's surface from the orbit. As the streams of impact debris landed on Earth, this material would be expected to be globally distributed, creating a uniform layer enriched to the same extent in siderophile elements, the class of elements to which iridium belongs. In other words, the mix of siderophile elements in the layer should match the mix of elements from where they originated.

The question was, did this mix match with what we could expect to find in an asteroid, or what we might find in the Earth's core? Munir and Steffanie observed the various siderophile elements did not occur in cosmic proportions within the El Kef boundary clay but resembled the relative abundances of siderophile elements in the Earth's crust. They also observed that the amount of enriched nickel and cobalt found in the samples were in Earth's crustal proportions, not the proportions found in meteorites.

Their analysis didn't stop there. Because Munir and Steffanie used a sensitive microanalytical instrument (that goes by the long name of "laser ablation inductively coupled plasma mass spectrometry"), they were able to measure the distribution of elements within mineral phases. They found that the siderophile elements were concentrated in goethite ($FeOOH$), a mineral that resembles rust, that was known to form by oxidation of a precursor mineral, pyrite ($FeS2$). This discovery was something of an "Aha!" moment. Pyrite is widely known to occur in KPB (boundary) clays, and forms in marine sediments when goethite from pyrite—after the sediment has been emplaced—provided a means by which trace metals could be highly concentrated in the KPB clay but not in the surrounding sediments.

But pyrite is not limited only to the KPB clay but occurs elsewhere in the El Kef sediments. Everywhere they found goethite formed from pyrite, they also found high concentrations of siderophile elements. The mystery was now cleared. Invoking an asteroidal impactor was unnecessary as

EXPLAINING THE "IRIDIUM ANOMALY" 237

the more detailed chemical data available no longer fit the hypothesis, but the presence of ubiquitous altered pyrite, which had now altered to goethite, provided a simple and efficient mechanism to highly concentrate siderophile elements from crustal sources both at the KPB and in other places within the column of sediments sampled at El Kef.

But perhaps I should let Munir explain. With his permission, I'm including this excerpt from his "confession" to me. Not only does it explain his scientific work elegantly and concisely, but it is also exemplary of the rigor and humility that characterizes the best of scientists.

I was an impactor when I first met Gerta Keller. For years, I had wanted to investigate the KPB-asteroid impact to learn about the abundances of the highly "iron-loving" siderophile elements, of which iridium is a member. These elements are most abundant in the Earth's metallic core but very rare in Earth's surface that forms the crust. An impactor (or bolide) that has a high abundance of iron and crashes into Earth adds a disproportionately larger amount of Ir and other highly siderophile elements to the crust. This will easily overwhelm the naturally low abundances in crustal rocks and leave a fingerprint of the impactor's unique formation history. The high Ir concentration in crustal rocks was the main argument that Alvarez used as proof of his impact theory.

It was no secret to me that in the 1980s the emphasis on iridium was because it was easy to measure by neutron activation analysis. This is a method that bombards the sample with neutrons and causes the elements present in the sample to form radioactive isotopes, which are easier to analyze than any aspect of iridium chemistry. But Ir enrichment alone was only part of the story. Other members of the highly siderophile elements are also enriched and could yield more complete information, including ruthenium and osmium, rhodium and iridium, palladium and platinum, and neighboring gold, which sit below the elements iron, cobalt, and nickel in the Periodic Table.

I knew it wasn't going to be simple—the existing measurements revealed the highly siderophile elements were all enriched, but

238 THE LAST EXTINCTION

that the patterns were messy looking and didn't resemble the expected chondritic pattern. A chondrite is a rock made up of tiny dust particles that formed from the disk of dust around the growing young star that became our Sun and from which the planets ultimately originated. Chondrites have elemental abundances in similar proportions to the composition of the Sun.

The messiness of the siderophile element patterns at the KPB never had a cosmic look to it. The KPB clay patterns varied from site to site. This suggested that after deposition, the siderophile elements had been mobilized by chemical changes (diagenesis), which convert soft, freshly deposited sediment into hard sedimentary rocks often over millions of years as the pore waters are gradually squeezed out of the compacting rock pile. To test this potential mechanism, the siderophile elements have to be examined not just at the KP boundary but also above and below to locate the elusive elements that could have migrated away from their original layer of deposition. Ideas I had plenty—samples none, till I met Gerta Keller.

When Gerta came to give her talk, I entered the room thinking, *What could she tell me that would change my mind that the siderophile elements were not from an impactor? What about the shocked quartz, the nickel-rich spinels, the impact melt spherules, the Chicxulub crater, the soot in the clays that evidenced a global wildfire at the KP boundary? How could she explain away evidence of a catastrophe that had accumulated into a seemingly invincible argument over four decades of research?*

I also had another line of work that concentrated on the chemical analysis of lavas. My co-worker, Shuying Yang, had shown that volcanoes emit copious amounts of metals into the atmosphere during eruptions, a fact not lost on volcanologists who studied the volatile emissions from erupting volcanoes. This was just as true of volcanic rocks from Mars as was from Earth, and the outgassed metals cool to form aerosols that are deposited in the sediments on the surface of the planet. When Gerta showed the link between extinction and volcanism, it occurred to me that we could search for evidence of volcanic outgassing coincidental with the extinction of planktonic forams.

Gerta generously provided us with samples from El Kef that spanned the KPB. My graduate student Steffanie took an interest in the project, and we analyzed the El Kef sediments by laser ablation inductively coupled plasma mass spectrometry, which permits analysis of about sixty elements simultaneously in a tiny volume of rock visible under a microscope. The laser cuts a track in the sample that is smaller than a hair's width and barely visible to the naked eye. Under the microscope, we could see iron-rich rusty bands that had accumulated huge amounts of many trace metals, including iridium. Many of the elements were not indicative of meteoritic debris. How did they accumulate with Ir along the same hairline iron-rich rusty bands?

Our first clue came from the high abundances of nickel and cobalt, which were in the same proportions as in the continental crust. If nickel and cobalt of terrestrial origin were highly concentrated into the rust layer, what about other elements? We took a critical look at the abundance pattern of the siderophile elements to see whether any had leaked out to the surrounding sediments above and below the boundary clay. The patterns stayed the same, even if the abundances were extremely high in the KPB rusty layer and lower elsewhere—there was no preferential loss of any element. The siderophile element pattern in the rusty band was distinctly like the pattern observed in continental crust, bearing little evidence of a cosmic origin.

Another clue came from osmium, which is a metal very similar in its cosmic abundance and behavior to iridium. The KPB sediments had a distinctly high ratio of osmium to iridium, which is unlike cosmic materials where the two elements occur in a rather constant ratio. But continental sediments gain more osmium than iridium from volcanic emissions. This explained the higher osmium concentrations and volcanic source.

Clues like this helped us home in on a troubling truth: The elemental abundance now showed a terrestrial pattern in the

240 THE LAST EXTINCTION

enriched metals, few of which could plausibly be argued as meteoritic infall. We were looking at something no geochemist had hoped to find: a mechanism by which iridium and other siderophile elements are enriched by processes that were clearly internal to that of the continental crust and required no meteoritic input. A little volcanism could help.

So, what was the elusive mechanism that had concentrated so many metals at the KPB red clay? We needed to understand what happens when water from the surface (with dissolved oxygen) travels through an ancient rock that contains iron-rich minerals. The mechanism that we inferred involves the original presence of iron-rich minerals that weathered in place to create the rusty bands. The rust from the corroding minerals acted as a trap for the very metals that we now find concentrated, and it could have concentrated such metals long after the mass extinction was over.

So, the iridium was not deposited from the fireball of an impact, but leaked in over millennia after the extinction—how ironic. The iridium is not part of a killer asteroid, but a passive interloper that accumulated long after the extinction during the weathering of the marine rocks at El Kef by oxygenated groundwaters. So, the ballistics experts got it all wrong by focusing on iridium and over-emphasizing the lack of mechanisms to accumulate large amounts of iridium in crustal rocks. Game over!

It's ironic that a former impactor toppled the last pillar of the impact theory: the Ir anomaly. Munir is the only impactor I've ever met with an open mind, willing to listen to evidence, and like a great science detective dig for truth and dare to follow it to its conclusion. It took me four decades to find such a scientist with the knowledge, guts, and courage. And now he has trained the next generation of young women scientists who are also courageous and open-minded to continue the battle for truth in science. How long will it be until the impactors have the guts, courage, and knowledge to recognize they've been wrong all along?

18

ARE WE LIVING *in the* SIXTH EXTINCTION?

FOR MANY YEARS, I'VE TAUGHT A COURSE AT PRINCETON ON "Catastrophes and Mass Extinctions." In this class, my students learned about the five major extinction events—ones that resulted in the death of 75 percent and up to 90 percent of living species—beginning with the Late Ordovician about 445 million years ago all the way through the fifth extinction, the Cretaceous–Paleogene, with which you are now very familiar. We discussed the immense volcanism that accompanied all five extinction periods, and the long-term climate warming and high-stress environments that caused species populations to dramatically decline as prelude to extinctions.

When we discuss the fifth mass extinction, we focus on how, for the last few thousand years before the end of this long period of intense volcanic activity, extinction rates and volcanism both sharply increase, as gigantic, pulsed Deccan eruptions created enormous greenhouse gas inputs that caused rapid warming of several degrees, and also poisoned the atmosphere with toxins resulting in acid rain and ocean acidification. Over a very short time, the stress to ecological systems reached the tipping point and life ended with rapid mass extinctions on land and in the oceans.

As the geological record shows, this type of climate change has happened repeatedly over Earth's long history, and we are seeing alarming

242 THE LAST EXTINCTION

signs of the pattern repeating itself today. It has gotten so bad, so quickly, that scientists have begun to wonder if we are in the midst of a sixth mass extinction right now. Of course, what differentiates this sixth ongoing mass extinction and rapid climate warming from the five that precede it is the absence of immense volcanic eruptions spewing out the deadly greenhouse gases. Instead, we humans managed to replace those natural events with our own homemade catastrophe that follows exactly the same well-worn path to Earth's previous mass extinctions.

It began with the rampant burning of coal, oil, and gas, releasing vast quantities of greenhouse gases into the atmosphere at rates twelve to sixteen times the peak volcanic eruptions during the dinosaur mass extinction. It has been exacerbated by the destruction of natural carbon sinks through rampant deforestation around the globe. And it has been further accelerated by the triggering of dozens of climactic feedback loops, like the melting of sunlight-reflecting ice at the poles, and the creation of ever more greenhouse gases from microbes released from thawing soil in the warming tundra regions of the Earth. The tipping point of rapid climate warming during the fifth mass extinction was reached around 4°C in the oceans and 7°C on land over a few thousand years.

At the rate we are going, we don't have a few thousand years to save our environment and ensure our survival. We don't even have a hundred years. The current rate of greenhouse gas input will likely more than double within a few years, due to population growth and the increasing adoption of energy-intense lifestyles by people around the globe. And we still must reckon with the thawing permafrost in Siberia, which will release colossal amounts of methane into the atmosphere. Methane is a greenhouse gas twenty-three times more potent than carbon dioxide, which has been the main source of climate warming so far. We are estimated to reach the 5°C tipping point for the sixth mass extinction as early as 2030 or as late as 2050. The humongous amount of methane released from the Siberian permafrost is likely to push us to the tipping point well before 2050.

Governments all over the world have repeatedly missed opportunities to mitigate the damage. The necessary steps to save our planet by reducing fossil fuel burning were proposed as far back as 1980, but ignored

ARE WE LIVING IN THE SIXTH EXTINCTION? 243

in favor of more pressing concerns, like economic expansion. By 2000, the threat was more apparent and imminent, but still little was done to reduce greenhouse gas input. The few rules that were in place have since been undone or mostly ignored. This blatant disregard for our future in favor of raking in profits reached its apex between 2016 and 2020 under the Trump administration—and continues today.

The direst prognosis—and a reasonable one from where we sit today—will see the extinction of about half the species on Earth since industrialization. Most of them are already endangered, their populations—and their habitats—permanently reduced by 40 to 60 percent. This includes large mammals, small vertebrates, reptiles, birds, bees, insects, and most of our sea life. It's a typical prelude to a mass extinction. Another 25 percent will disappear rapidly once the tipping point is reached. It's no secret that human populations are driving these extinctions, and that our descendants will suffer as the worst effects emerge. Changing weather patterns and ecological disruption will affect our ability to farm and fish, which will result in our own population decrease and likely extinction because life, as we know it, has become unsustainable. Sometimes I imagine a scientist in the distant future, sifting through Earth's sediment layers and finding the record of our history, written in stone and fossils and bits of degrading plastic, a mere blip in the geological record known as the Man-Made Mass Extinction.

At this point, the best-case scenario is that the ruling governments and mega-rich of the world will eventually come to their senses and realize that their own futures depend on humanity's survival and find a way to manage our resources wisely and for the benefit of all. Even then, our near future is going to be fraught, likely filled with increasingly violent weather events, wars over resources, displaced populations, mass starvation, and declining quality of life.

We are well past the middle of the sixth mass extinction. It's a frightening statement. The mere thought conjures up the dinosaurs in their last moments, writhing and succumbing to volcanism's heat and toxic gases. Unlike the dinosaurs, we have the ability to understand that we are now staring into that same abyss, and that there is little time left. Perhaps in that self-awareness there is some hope.

244 THE LAST EXTINCTION

I've spent forty years fighting for truth in the Dinosaur Wars. It was an epic journey of many ups and downs. I love my research, the highs of exhilarating discoveries, the joys of teaching and training students, and meeting wonderful friends and collaborators. My most positive influence are young scientists who increasingly pursue climate change as the most likely cause of all major mass extinctions and catastrophes in Earth's history. They nurture and improve our planet Earth's survival for all living creatures. It's the young who are giving us hope, even though we have already passed the eleventh hour and the clock is ticking toward the point of no return.

I plan to continue my research career until we know if these dire predictions are right. Until then, I will celebrate the discoveries and achievements of the next generation of young and future scientists who inspire us to save our planet. They are the new truth seekers. I'm confident they can win as the old guard disappears like the dinosaurs. And I may be the last one.

ACKNOWLEDGMENTS

It's not easy writing a book, especially one that spans nearly a lifetime in the academic wilderness of asteroid theories, impacts, mass extinctions, evolution of life, Mexico's Chicxulub impact, Deccan volcanism, and climate change. I began my science life totally unexpectedly at age twenty-three on a stopover in San Francisco. Although I craved to learn about science, this was not permissible for poor Swiss schoolgirls. The best they could achieve was maid, salesgirl, or seamstress. If that wasn't enough for me, I would be sent to the nunnery in total isolation. I escaped this Swiss life at age seventeen and traveled through the world's most dangerous wars to experience life, until I'd seen enough.

San Francisco was the last stage of my journey, and it was weirder than anything I had ever experienced. It was the time of flower children and the Free Speech Movement—when Haight-Ashbury was overflowing with hippies, countless drugs, and great songs. I documented much of it on camera; it was all so crazy. One day, I remember the hippies asking me about my studies, to which I confessed I had none and no money to pursue such a thing. "Go to San Francisco's City College," they told me. "You don't need money as long as you've got good grades." Well, I went and took the entrance exam, and from that day forward, I never lacked scholarship funds. My first thank-you is to those hippies, who encouraged me and allowed me to begin my academic journey.

Next, I want to thank my teachers from San Francisco City College and San Francisco State University, including Professors Ray Sullivan and Hans Thalmann. I also want to thank Professor Jim Ingle of Stanford University for being such a fantastic teacher, and Professor William (Bill) Evitt for helping increase my understanding of the microfossil world and for introducing me to his former PhD advisor, Dewey McLean.

246 ACKNOWLEDGMENTS

One of my oldest colleagues and friends is William (Bill) MacDonald, whom I first met at the NSF conference in Jamaica. I remember I had recently accepted associate professorship at Princeton. During that conference, Bill was great company, and every day, he joined my runs. We talked and talked, though he frequently took breaks to catch his breath. I only learned later that Bill was not, in fact, a runner, but decided to join me on my morning jogs through "Bunny Bay" for my own safety. He has been a wonderful friend to me, and I am so grateful to him.

Another thank-you goes to one of my long-term collaborators, Thierry Adatte from the University of Lausanne in Switzerland. To this day, I can't shake the memory of our crazy sightseeing day in the Atlas Mountains of Morocco. Though our small rental car was not suitable for off-road driving, all started out well as we searched for a beautiful gorge that Thierry recalled from a teenage visit to Morocco. Finally, after five hours of no signs of civilization, and approaching darkness, we found not the gorge but instead a valley with a minaret and a dozen people watching our small car approach in silence. Without stopping, we found a small road and hightailed it out the valley. It was sheer lunacy. We never talked about it again.

I would also like to thank Wolfgang Stinnesbeck from Linares, Mexico, another long-term collaborator and friend who often joined me leading Princeton University field trips in Mexico. I recall how, one year, Wolfgang proposed the field trip to the smoldering underground fires in northern Mexico, which I hesitantly agreed to. As we stood on the road near the fires, Wolfgang made joke after joke about the fires as he lectured. Suddenly, the 6'4" man disappeared up to his neck, desperately holding on to his head. Of course, everyone continued to laugh, believing it was another joke. But it was not. Wolfgang had fallen into the fire. Somehow, he managed to save himself, pants ablaze, but no longer joking.

It was not only the unpredictable danger that made geology so much fun, but also the incredible people I got to experience these types of stories with, so thank you to Wolfgang and Thierry for being with me through it all.

ACKNOWLEDGMENTS 247

In addition, I've had the incredible opportunity to work with some amazing scientists over the course of my career, and I want to spend time thanking them as well.

To Zsolt Berner, PhD, isotope geochemist at the University of Karlsruhe, Germany: You are one of the very best scientists I've collaborated with for decades. I'm truly grateful for the many years we've had the chance to work together and for our many discoveries and results which continue to prevail. I wish you all the best and good health in your retirement.

To Jorge Spangenberg, PhD, isotope geochemist at the University of Lausanne, Switzerland: You are a fantastic isotope geochemist who leaves no questions about accuracy. I have been very lucky to work with you, after Zsolt's retirement. Thank you, Jorge. It's been a great pleasure.

To the women scientists who have paved similar, impossible paths through a man's world—specifically to Doris Stüben, PhD, geochemist and first leading woman scientist at the University of Karlsruhe, and to Nallamuthu Malarkodi, who set an incredible example of extreme perseverance after decades of survival in the trenches of prejudice. Thank you both for being inspirations and for leading the way for the rest of us.

During the pandemic, when most research was shut down at Princeton, undergraduate student Udit Basu worked on my mercury project in Mexico where we discovered mercury fallout that predated the Chicxulub impact by 200,000 years. Thank you, Udit. Discoveries like these make all the difference in science.

My next thank-you goes to geophysicist Vincent Courtillot for working many years on the Deccan volcanism in India and searching for the KPB mass extinction. Also, thank you to my former PhD students Jahnavi Punekar (IITB) and Paula Mateo (Caltech) for your dedication and hard work.

I would also like to thank Ted Nield and Andrew Kerr for the immense support and advice you both gave me during the ugliest parts of the Dinosaur Wars.

I'm deeply grateful to astrophysicist and geochemist Munir Humayun for being willing to investigate your own trust in the impact theory and its plausibility. You and your graduate student Steffanie Sillitoe-Kukas

248 ACKNOWLEDGMENTS

brought down the last pillar of the impact theory. It is one of the greatest stories of research and discoveries I've seen in my time.

My gratitude goes to many of my colleagues working on Deccan volcanism, including Blair Schoene, Kyle Samperton, Mike Eddy, and our field guide Syed Khadry.

I'm thankful to Norman McLeod, both colleague and friend. I still think of that make-believe conversation—it is the funniest story.

I'm grateful to Rex Dalton, investigative reporter for Nature, who searched the records of the Chicxulub Impact Wars in 2003 and helped uncover the fabrication therein. You were willing to risk your job in the search for truth. Thank you, Rex, you are a hero.

Thank you, Brian Holland, for many fun trips to Belize with Andy Majda. Though I can't quite thank you for the "fun" we had on our last boat adventure, which was eventually shot full of bullet holes by a rival drug boat waiting for revenge. I can say, without a doubt, it was my most harrowing trip at sea, and that it was the last time I'll ever do fieldwork in Belize.

My gratitude also goes to Boualem Khouider, for your great company during Andy Majda's last journey to the Swiss Alps, with Roger, Caludio, and Ernst. It was a wonderful journey--one I will never forget.

I'm forever grateful to Elizabeth Evans, who taught me how to simplify my scientific language and write great stories that are accessible for scientists and nonscientists alike. Most of all, thank you for your unwavering friendship through many years of the Dinosaur Wars.

A special thank-you goes to Jeff Alexander, who also taught me how to compose my scientific work: beautifully, straightforward, and always with the right punch when needed. I couldn't have gotten here without you.

Finally, I would like to thank my Geosciences Department, specifically Nora Zelizer, Mike Morris, Georgette Chalker, Mary Rose Russo, Eva Groves, Sheryl Robas, Doreen Sullivan, Mae Castro, and so many others I didn't have the space to name. What an up-and-down ride this has been. You are all like family to me.

Thank you to everyone who has been a part of this journey with me, and to everyone who has decided to pick up this book and read it.

I hope my journey inspires you to stand firm in what you believe and to push back, even when it seems like the world is against you. And I hope you find incredible people like I have to help make the road more bearable along the way. I am incredibly grateful for each and every one of them.

NOTES

1 Around 2005, a change in the classification system replaced Tertiary for Paleogene (P or Pg). For consistency in this book, I will update the KTB mass extinction boundary to KPB.

2 Ian Warden, *The Canberra Times*, May 20, 1984.

3 An asteroid is a giant body with known sizes up to 940 kilometers in diameter (Ceres asteroid) and tends to break up into smaller pieces known as meteorites; the one that hit Earth was about 10 kilometers in diameter and is a meteorite. Comets are also very large, often consist of ice and rock, and break apart, forming comet showers.

4 Browne, Malcolm W., "The Debate Over Dinosaur Extinctions Takes an Unusually Rancorous Turn." *New York Times*, January 19, 1988.

5 Luis Alvarez Obituary, Sept. 2, 1988, *Chicago Tribune*.

6 Alvarez L.W. et al., "Search for Hidden Chambers in the Pyramids," *Science*, 1970, v. 167, pp 832–39.

7 Alvarez's 1982 lecture was published in 1983 in Proc. Natl. Acad. Sci. USA from where quotes are taken.

8 David M. Raup, 1987. *The Nemesis Affair, A Story of the Death of the Dinosaurs and the Ways of Science.*

9 To this day, science discriminates against articles that do not support the impact theory.

10 Browne, Malcolm W., "The Debate Over Dinosaur Extinctions Takes an Unusually Rancorous Turn." *New York Times*, January 19, 1988.

11 Dewey M. McLean, "A Terminal Mesozoic 'Greenhouse': Lessons from the Past," *Science*, August 4, 1978: vol. 201, Issue 4354, pp.401-406. DOI: 10.1126/science.201.4354.401.

12 Browne, Malcolm W., "The Debate Over Dinosaur Extinction Takes an Unusually Rancorous Turn," *New York Times*, January 19, 1988.

13 Dewey's webpage disappeared after his death, but he sent many parts of it to Gerta Keller.

14 Ibid.

15 Alvarez W., Asaro, F., Michel H., and Alvarez L.W. 1982. Iridium anomaly approximately synchronous with terminal Eocene extinctions. *Science*, May 21, 1982: 216 (4548), 886–88, https://doi.org/10.1126/science.216.4548.886.

16 Gerta Keller, Steve D'Hondt, and Tracy L. Vallier, 1983. "Multiple Microtektite Horizons in Upper Eocene Marine Sediments: No Evidence for Mass Extinctions." *Science* v. 221, No. 4606 (July 8, 1983) http:/www.jstor.org /stable/1691574.

252 NOTES

17 Author correspondence with the late Doug Nichols of the Denver Museum of Nature & Science.

18 Shortly afterward, Jason Morgan, Princeton University, convinced Courtillot that Kirschvink's review was biased; they agreed to review this paper in the fall of 2003 in PNAS, which *Nature* had rejected as "unsuitable." In 2009, Howard Falcon-Lang, Royal Holloway University, London, (editor: J. Geological Society of London) rejected our paper based on Smit and Schulte, which they had already reviewed and rejected by three different reviewers and rejected each time by the same impactors. This time, I wrote a letter to Falcon-Lang and explained what was happening to this paper. I received a letter back saying he reviewed the paper himself and it is an important contribution, needing some clarification for publication. We finally won this one.

19 The Alps run throughout Switzerland, and the Swiss commonly refer to them as either the "pre-Alps," an older, smaller range, or the "high-Alps," the newer, taller range.

20 The village church was Protestant and my family was Catholic, so we did not attend. According to the village register, Frau Babaly's last name was Tinner-Berger. Her husband died in an accident at an early age. She never remarried.

21 The impact spherules were first reported by Hildebrand and Boynton in an abstract at the Lunar and Planetary Conference in March 1990. In September 1990, Glen Izett, Florentin Maurrasse, and others visited this Haiti locality to collect and study impact spherules, which they reported by late fall in a US Geological Survey Open-File Report in 1990.

22 Hildebrand, A.R., Penfield, G. T. , Kring, D. A. , Pilkington, M., Camargo, A., Jacobsen, S.B., Boynton, W. 1991, "Chicxulub Crater: A Possible Cretaceous/Tertiary Boundary Impact Crater on the Yucatan Peninsula, Mexico": *Geology*, v. 19, p. 867–71.

23 Penfield, G. T. , and Camargo Z., A., 1981, definition of a major igneous zone in the central Yucatán platform with aeromagnetics and gravity, in technical program, abstracts, and biographies (Society of Exploration Geophysicists 51st Annual International Meeting): Los Angeles, Society of Exploration Geophysicists, p. 37.

24 Hildebrand et al., 1991, "Chicxulub Crater: A Possible Cretaceous/Tertiary Boundary Impact Crater on the Yucatan Peninsula, Mexico." *Geology*, v. 19, p. 867–71.

25 Smit, J., Montanari, A., Alvarez, W., Hildebrand, A. R., Margolis, S. V., Claeys, P., Lowrie, W., Asaro, F., 1992. Tektite-bearing, deep-water clastic unit at the Cretaceous-Tertiary boundary in northeastern Mexico. *Geology* v. 20, 99–103.

  Keller, G., 1989b. Extended K/T boundary extinctions and delayed populational change in planktic foraminiferal faunas from Brazos River, Texas, Paleoceanography, 4(3): 287–332.

  Keller, G., 1991. Extended period of extinctions across the Cretaceous/Tertiary Boundary: reply to comments by T. Hansen and J. Bourgeois, Geol. Soc. America Bull., p. 435.

  Stinnesbeck, W., Barbarin, J. M., Keller, G., Lopez-Oliva, J. G., Pivnik, D. A., Lyons, J. B., Officer, C. B., Adatte, T., Graup, G., Rocchia, R. and Robin, E., 1993. "Deposition of near K/T Boundary clastic sediments in northeastern Mexico: Impact or Turbidite Deposits?" *Geology*, v. 21, p. 797–800.

26 Stinnesbeck, W., Barbarin, J. M., Keller, G., Lopez-Oliva, J. G. , Pivnik, D. A., Lyons, J. B., Officer, C. B. , Adatte, T., Graup, G., Rocchia, R. and Robin, E., 1993. "Deposition of near K/T Boundary clastic sediments in northeastern Mexico: Impact or Turbidite Deposits?" *Geology*, v. 21, p. 797–800.

27 Arthur A. Meyerhoff (1928–1994) was a consultant for PEMEX examining the crater cores on Yucatan between 1976 and 1978. He determined the limestone overlying the breccia was the latest Cretaceous in age based on forams. I discussed the cores with Meyerhoff in 1992–1993 and saw no reason to doubt him.

28 Conway, Erik M., Donald K. Yeomans and Meg Rosenburg, A History of Near-Earth Objects Research, NASA Department of Communications, pp. 105–6, Chapter 5. https://www.nasa.gov/history/history-publications-and-resources /nasa-history-series/a-history-of-near-earth-objects-research/.

29 Kerr, R. A., 1994. Testing an Ancient Impact's Punch. *Science* vol. 5152 (March 11, 1994), pp.1371–72.

30 We first presented this study on the primary impact spherule layer in the Geological Society of America Special Paper 356, 2002 (Keller, G., Adatte, T., Stinnesbeck, W., Affolter, M., Schilli, L., and Lopez-Oliva, J. G.), 2002. Multiple spherule layers in the late Maastrichtian of northeastern Mexico, In Koeberl, C. and MacLeod, K.G., eds., Catastrophic Events and Mass Extinctions: Impacts and Beyond: Boulder, Colorado, Geological Society of America Special Paper 356, p.145–61.

31 Deccan dating at the time was 300,000 years and subsequently updated to reveal 200,000 years pre-KPB.

32 Keller, G., Adatte, T., Berner, Z., Pardo, A., Lopez-Oliva, L., 2009. New Evidence Concerning the Age and Biotic Effects of the Chicxulub Impact in Mexico. J. Geol. Society, London 166, 393-411. doi:10.1144/0016-76492008 116.

33 http://www.sciencedaily.com/releases/2007/10/071029134743.htm.

34 Bhandari et al., 1993, 1994; Garg et al., 1996.

35 *Science* commentary by Richard Kerr, 2019 (same page as Schoene et al. and Sprain et al., 2019).

36 Letters to *Science* in critique of Schulte et al., 2010. May 20, 2010. www. sciencemag.org: Archibald, et al. (thirty co-authors), "Cretaceous Extinctions: Multiple Hypotheses"; Courtillot and Fluteau, "Cretaceous Extinctions: The Volcanic Hypothesis"; Keller et al. (six co-authors) "Cretaceous Extinctions: Evidence Overlooked."

37 Vincent Courtillot and Frédéric Fluteau, 2010. "Cretaceous Extinctions: The Volcanic Hypothesis". Institut de Physique du Globe, Paris, France. *Science*, May 21, 2010, 238, p. 973. www.sciencemag.org.

38 Jay Melosh died September 11, 2020.

39 The Pew Research Center's ongoing survey of Americans' trust in the government, and in each other, has shown a steady decline since it began in 2014. Compared to similar surveys from decades ago, the decline in trust is even more pronounced. https://www.pewresearch.org/politics/2019/07/22 /americans-struggles-with-truth-accuracy-and-accountability/.

ABOUT THE AUTHOR

GERTA KELLER is a professor of paleontology and geology in the Department of Geosciences at Princeton University. She has published more than 260 articles in international journals and is a leading authority on catastrophes, mass extinctions, and the biotic and environmental effects of impacts and volcanism. She has co-authored five academic books, and her work has been featured in TV documentaries and news features, radio and podcast interviews, and print and web media, most notably in a widely circulated profile in *The Atlantic*.

For full color graphics, visit
https://diversionbooks.com/color-graphics-the-last-extinction/
or scan the QR code below.